D1743416

GCSE 9-1

PHYSICS

REVISION GUIDE

DAN FOULDER

About this Book

Whether you like to learn by seeing, hearing or doing (or a mix of all three), this unique guide uses three different approaches to help you engage with each topic and revise in a way that suits you. It will reinforce all the skills and concepts on the GCSE 9-1 courses for AQA, Edexcel and OCR Gateway, ensuring you are well prepared for your exams.

Key Features

This guide neatly packages the GCSE course into short revision modules to make planning easy.

- Striking page designs, images and diagrams help you engage with the topics.
- Hands-on revision activities.
- Download the relevant track for an audio walk-through of each module.
- Simple, concise explanations for effective revision.
- Keyword boxes to build vocabulary.
- Quick tests to check your understanding of each module.
- Mind maps summarise the key concepts at the end of each topic and show how they are linked.
- Exam practice questions test how well you have understood the topics. Answers are provided at the back of the book.
- **HT** Higher tier information highlighted.
- **WS** Working scientifically covers practical skills and data-related concepts.

Letts GCSE 9-1 Physics Exam Practice Workbook (ISBN 9780008318383) provides further practice.

Download the FREE audio book from lettsrevision.co.uk/GCSE

Contents

Forces

Forces

Scalar quantities have magnitude only.

Vector quantities have magnitude and an associated direction.

A vector quantity can be represented by an arrow. The length of the arrow represents the magnitude, and the direction of the arrow represents the direction of the vector quantity.

Contact and non-contact forces

A force is a push or pull that acts on an object due to the interaction with another object. Force is a vector quantity. All forces between objects are either:

- contact forces – the objects are physically touching, for example:
  friction, air resistance, tension and normal contact force

  or

- non-contact forces – the objects are physically separated, for example:
  gravitational force, electrostatic force and magnetic force.

Friction

Air resistance

Gravity

- **Weight** is the force acting on an object due to gravity.
- All matter has a gravitational field that causes attraction. The field strength is much greater for massive objects.
- The force of gravity close to the Earth is due to the gravitational field around the Earth.
- The weight of an object depends on the gravitational field strength at the point where the object is.
- The weight of an object and the mass of an object are directly proportional (weight $\propto$ mass). Weight is a vector quantity as it has a magnitude and a direction. Mass is a scalar quantity as it only has a magnitude.

Weight is measured using a calibrated spring-balance – a **newtonmeter**.

Weight can be calculated using the following equation:

weight = mass × gravitational field strength
$$W = mg$$
- weight, W, in newtons, N
- mass, m, in kilograms, kg
- gravitational field strength, g, in newtons per kilogram, N/kg

Example:
What is the weight of an object with a mass of 54 kg in a gravitational field strength of 10 N/kg?

weight = mass × gravitational field strength
$$= 54 \text{ kg} \times 10 \text{ N/kg}$$
$$= 540 \text{ N}$$

Resultant forces

The resultant force equals the total effect of all the different forces acting on an object.

HT

20 kN

When the truck is accelerating the forces on it are unbalanced

10 kN | 1 kN

The resultant force is therefore 9 kN

20 kN

9 kN

If the forces on an object are balanced, the resultant force is zero. If the object is stationary it remains stationary and if it is moving it continues moving at a constant speed.

A single force can be resolved into two components acting at right angles to each other. The two component forces together have the same effect as the single force.

Keywords

Scalar ➤ A quantity that only has a magnitude

Vector ➤ A quantity that has both a magnitude and a direction

Weight ➤ Force acting on an object due to gravity

Work done and energy transfer

Work is done when a force causes an object to move. The force causes a displacement.

The work done by a force on an object can be calculated using the following equation:

> **work done = force × distance moved along the line of action of the force**
>
> $$W = Fs$$
>
> ➤ work done, W, in joules, J
> ➤ force, F, in newtons, N
> ➤ distance, s, in metres, m (s represents displacement, commonly called distance)

Example:

What work is done when a force of 90 N moves an object 14 m?

work done = force × distance moved along the line of action of the force

work done = 90 × 14 = 1260 J

One joule of work is done when a force of one newton causes a displacement of one metre.

1 joule = 1 newton metre

Work done against the frictional forces acting on an object causes a rise in the temperature of the object.

Draw two more large diagrams of the truck shown at the top of the page. Alter the force arrows to represent the truck decelerating and the truck stationary. **HT**

1. What is the difference between a vector quantity and a scalar quantity?
2. Give two examples of contact forces.
3. What is the weight of a 67 g object in a gravitational field strength of 10 N/kg?
4. What is 78 Nm in joules?

Forces and elasticity

Inelastically deformed

Elastically deformed

When an object is stretched and returns to its original length after the force is removed, it is **elastically deformed**.

When an object does not return to its original length after the force has been removed, it is **inelastically deformed**. This is **plastic deformation**.

Extension

The extension of an elastic object, such as a spring, is directly proportional to the force applied (extension ∝ force applied), if the limit of proportionality is not exceeded.

➤ **Stretching** – when a spring is stretched, the force pulling it exceeds the force of the spring.
➤ **Bending** – when a shelf bends under the weight of too many books, the downward force is from the weight of the books, and the shelf is resisting this weight.
➤ **Compressing** – when a car goes over a bump in the road, the force upwards is opposed by the springs in the car's suspension.

In order for any of these processes to occur, there must be more than one force applied to the object to bring about the change.

The force on a spring can be calculated using the following equation:

force = spring constant × extension
$$F = ke$$
➤ force, F, in newtons, N
➤ spring constant, k, in newtons per metre, N/m
➤ extension, e, in metres, m

Example:
A spring with spring constant 35 N/m is extended by 0.3 m. What is the force on the spring?

$F = ke$
$\quad = 35 \times 0.3$
$\quad = 10.5$ N

This relationship also applies to the compression of an elastic object, where the extension e would be the compression of the object.

A force that stretches (or compresses) a spring does **work** and **elastic potential energy** is stored in the spring. Provided the spring does not go past the limit of proportionality, the work done on the spring equals the stored elastic potential energy.

Spring balance

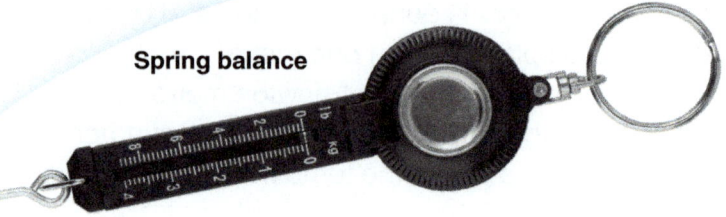

Force and extension

Force and extension have a linear relationship.

If force and extension are plotted on a graph, the points can be connected with a straight line (see Graph 1).

The points in a non-linear relationship cannot be connected by a straight line (see Graph 2).

Keywords

Elastically deformed ➤ Stretched object that returns to its original length after the force is removed
Inelastically deformed ➤ Stretched object that does not return to its original length after the force is removed

Graph 1 – Linear relationship

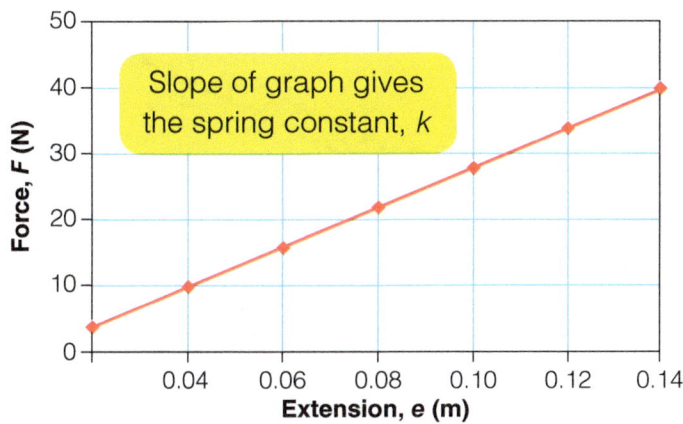

Slope of graph gives the spring constant, k

Graph 2 – Non-linear relationship

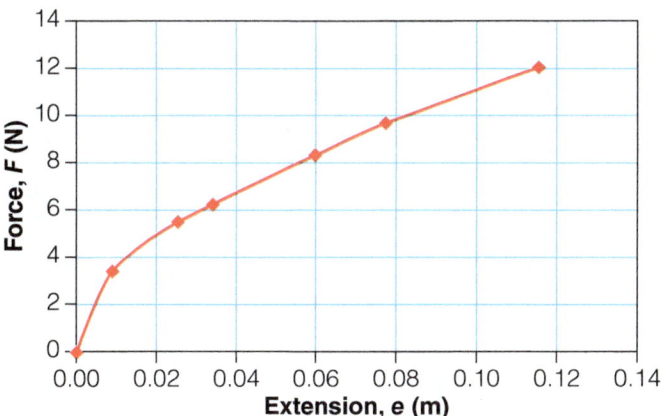

In order to calculate the spring constant in Graph 1, take two points and apply the equation.

$$\text{So, spring constant} = \frac{\text{force}}{\text{extension}}$$

$$= \frac{10}{0.04}$$

$$= 250 \text{ N/m}$$

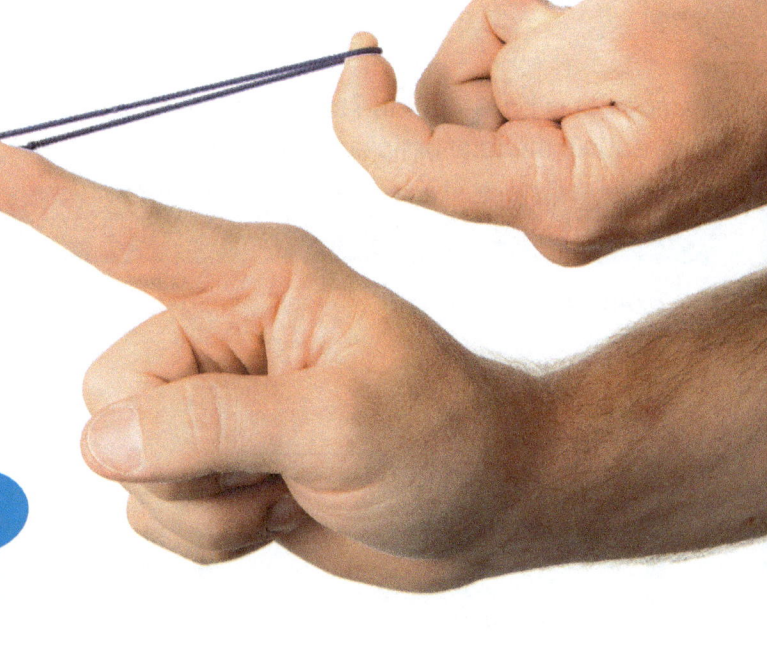

Attach a number of objects of different masses to an elastic band and measure how far the band extends. Repeat the investigation again using an plastic bag instead of a rubber band. How do your results compare?

1. What is the force of a spring with an extension of 0.1 m and a spring constant of 2 N/m?
2. When is the extension of an elastic object, such as a spring, directly proportional to the force applied?
3. What type of energy is stored in a spring?

2

Moments, levers and gears

Moments

The turning effect of a force is called the **moment** of the force.

The moment of a force is given by the following equation:

> moment of a force = force × distance to pivot
>
> $$M = Fd$$
>
> ➤ moment of a force, M, in newton metres, Nm
> ➤ force, F, in newtons, N
> ➤ distance to pivot, d, is the perpendicular distance from the pivot to the line of action of the force, in metres, m

> **Example:**
> A 30 cm spanner is used with a force of 36 N. What is the moment?
>
> $M = Fd$
> $\quad = 36 \times 0.3$
> $\quad = 10.8$ Nm

If an object is balanced, the total clockwise moment about a pivot equals the total anticlockwise moment about that pivot.

> the sum of clockwise moments = the sum of anticlockwise moments

For example:

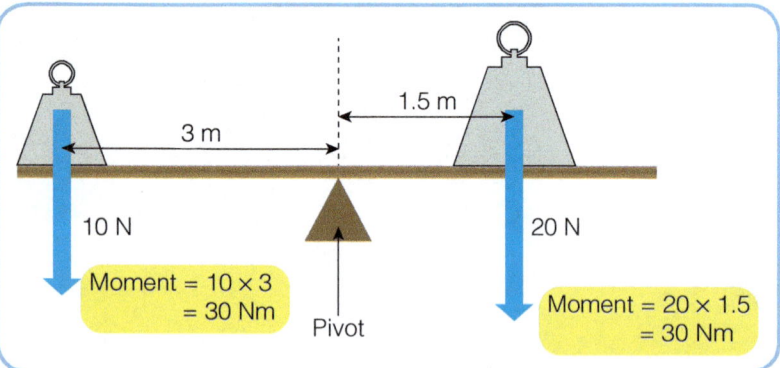

1.5 m

3 m

10 N

20 N

Moment = 10 × 3
= 30 Nm

Moment = 20 × 1.5
= 30 Nm

Pivot

Keyword

Moment ➤ Turning effect of a force

Levers

A lever reduces the force needed to carry out an action. The force is exerted around a pivot.

Levers and gear systems can both be used to transmit the rotational effects of forces. Levers and gears act as force multipliers, exerting a larger force than is being exerted on them.

For example:

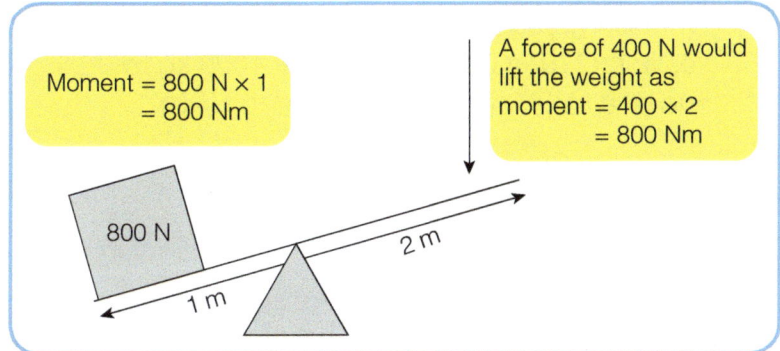

Moment = 800 N × 1
= 800 Nm

A force of 400 N would lift the weight as moment = 400 × 2
= 800 Nm

800 N

1 m

2 m

A **gear system** can also be used to transfer the rotational effect of forces. The teeth of gears interlock so when the first gear in a system rotates, it rotates the other gears that it is interlocked with. Two interconnected gears will rotate in opposite directions.

If gears are different sizes, the smaller gear in a system will rotate faster than the larger gear.

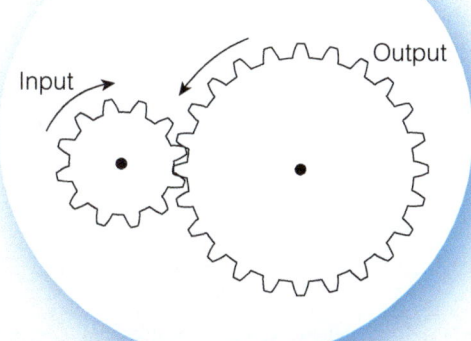

Input

Output

Close a door by pushing it at 10 cm intervals from the edge of the door towards the hinge. What do you notice about the force required as you move from the edge of the door towards the hinge?

1. What is a moment?
2. What statement can be made about the moments on a balanced object?
3. What is the moment when a force of 72 N is applied with a 70 cm lever?

Pressure in a fluid

Keyword

HT **Upthrust** ➤ Upward force that a liquid or gas exerts on a body in a fluid floating in it

Fluids

Liquids and gases are **fluids**.

The pressure in a fluid is caused by the fluid and the surrounding atmospheric pressure. The pressure causes a force to act at right angles (normal) to a surface. The pressure exerted on a surface by a fluid can be calculated using the following equation:

$$\text{pressure} = \frac{\text{force normal to a surface}}{\text{area of that surface}}$$

$$p = \frac{F}{A}$$

➤ pressure, p, in pascals, Pa
➤ force, F, in newtons, N
➤ area, A, in metres squared, m^2

Example:
What is the pressure of a force of 650 N over an area of 0.50 m^2?

$$\text{pressure} = \frac{\text{force normal to a surface}}{\text{area of that surface}}$$

$$= \frac{650}{0.5}$$

$$= 1300 \text{ Pa}$$

Liquids are relatively incompressible and the pressure in liquid is transmitted equally in all directions, meaning hydraulic systems containing liquids can be used to transmit pressure and so magnify a force.

HT The pressure exerted by a column of liquid can be calculated using the following equation:

$$\text{pressure} = \text{height of the column} \times \text{density of the liquid} \times \text{gravitational field strength}$$

$$p = h\rho g$$

➤ pressure, p, in pascals, Pa
➤ height of the column, h, in metres, m
➤ density, ρ, in kilograms per metre cubed, kg/m^3
➤ gravitational field strength, g, in newtons per kilogram, N/kg

Example:
What is the pressure at a depth of 40 m in seawater? (density of seawater = 1029 kg/m^3 and gravitational field strength = 10 N/kg)

$$\text{pressure} = 40 \times 1029 \times 10$$

$$= 411.6 \text{ kPa}$$

Pressure increases in a column of liquid as there are more particles applying the pressure. If the height of the column increases or the density of the liquid increases, this causes an increase in the number of particles and so increases the pressure.

HT Upthrust

A submerged object experiences a greater pressure on the bottom surface than on the top surface. This creates a resultant force upwards. This is **upthrust**.

An object less dense than the surrounding liquid displaces a volume of liquid equal to its own weight. This object floats as its weight is equal to the upthrust.

An object more dense than the surrounding liquid is unable to displace a volume of liquid equal to its own weight. This object sinks as its weight is greater than the upthrust.

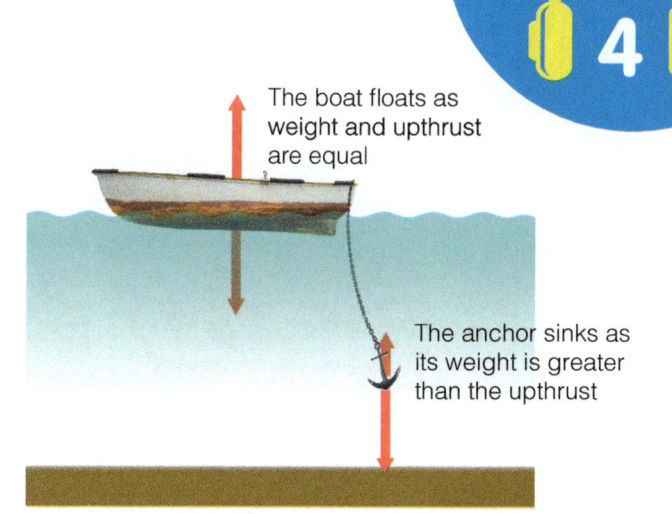

The boat floats as weight and upthrust are equal

The anchor sinks as its weight is greater than the upthrust

Atmospheric pressure

The atmosphere is a thin layer (relative to the size of the Earth) of air round the Earth. The atmosphere gets less dense with increasing altitude. The air molecules colliding with a surface create atmospheric pressure.

As height of an object increases, there are fewer air molecules above the object so their total weight is smaller. This leads to a lower atmospheric pressure.

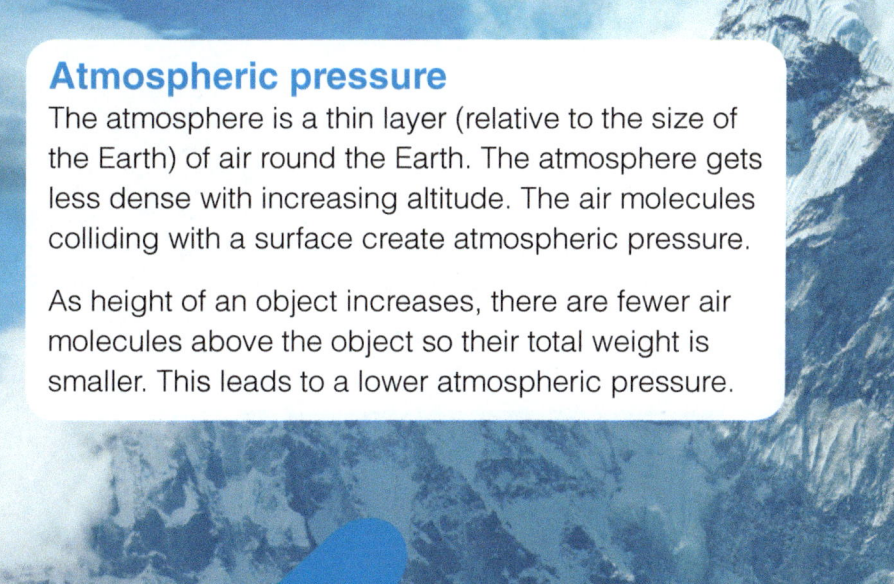

Carry out an investigation using a pin. First, place the pin onto some water in a glass and observe what happens. Then, put the pin on top of a small piece of paper and place the paper and pin onto the water. Observe the results.

HT **1.** What is the pressure in a 13 m column of water? (density of water = 1000 kg/m³ and gravitational field strength = 10 N/kg)

HT **2.** Explain, in terms of forces, why an object sinks.

3. Why does air pressure decrease with height?

Speed and velocity

5

Distance and speed

Distance is how far an object moves. As distance does not involve direction, it is a scalar quantity.

Displacement includes both the distance an object moves, measured in a straight line from the start point to the finish point, and the direction of that straight line. As displacement has magnitude and a direction, it is a vector quantity.

The speed of a moving object is rarely constant. When people walk, run or travel in a car, their speed is constantly changing.

The speed that a person can walk, run or cycle depends on:

- Age
- Terrain
- Speed
- Fitness
- Distance travelled

Some typical speeds are:
- walking – 1.5 m/s
- running – 3 m/s
- cycling – 6 m/s
- car – 20 m/s
- train – 35 m/s

The speed of sound and the speed of wind also vary.

A typical value for the speed of sound in air is 330 m/s. A gale force wind is one with a speed above 14 m/s.

Speed can be investigated in labs using equipment such as light gates.

For an object travelling at a constant speed, distance travelled can be calculated by the following equation:

> **distance travelled = speed × time**
> $$s = vt$$
> - distance, s, in metres, m
> - speed, v, in metres per second, m/s
> - time, t, in seconds, s

Example:
What distance is covered by a runner with a speed of 3 m/s in 6000 seconds?

$$\begin{aligned} \text{distance travelled} &= \text{speed} \times \text{time} \\ &= 3 \times 6000 \\ &= 18\,000 \text{ m} \\ &= 1.8 \text{ km} \end{aligned}$$

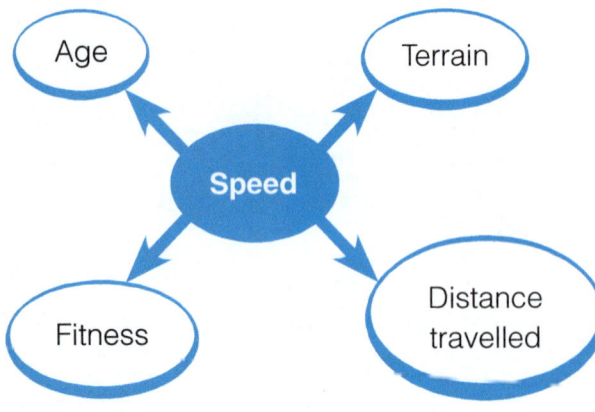

Velocity

The **velocity** of an object is its speed in a given direction. As velocity has a magnitude and a direction, it is a vector quantity.

Keywords

Displacement ➤ Distance travelled in a given direction

Velocity ➤ Speed in a given direction

HT A car travelling at a constant speed around a bend would have a varying velocity because it is changing direction.

When an object moves in a circle, the direction of the object is continually changing. This means that an object moving in a circle at constant speed, such as an orbiting satellite, has a continually changing velocity. For motion in a circle, there is a resultant centripetal force that acts towards the centre of the circle.

Centripetal force

Velocity

Time how long it takes to travel to school. Use a mapping website to calculate the distance. Now calculate the average speed of your journey. Replicate this investigation on different days. Is there much variation in your average speed?

1. Why is distance a scalar quantity?
2. What is the speed of a walker who covers 10 km in 2.5 hours?
3. What is the difference between speed and velocity?

Distance–time and velocity–time graphs

6

Distance and time

The distance an object moves in a straight line can be represented by a distance–time graph.

The speed of an object can be calculated from the gradient of its distance–time graph.

The graph below shows a person jogging.

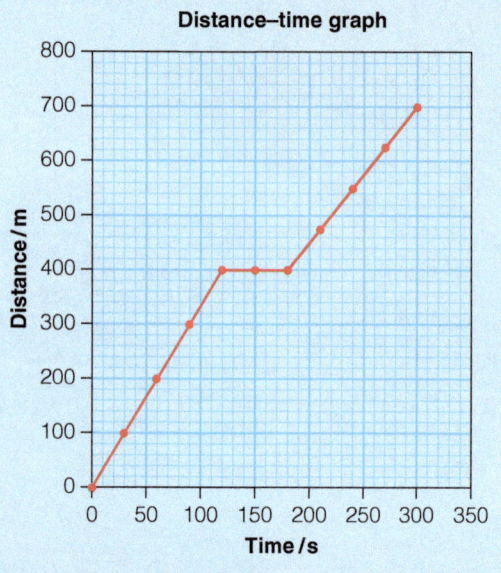

Distance–time graph

The average speed of this jogger between 0–120 seconds can be worked out as follows:

$$\text{speed} = \frac{\text{distance}}{\text{time}} = \frac{400}{120} = 3.33 \text{ m/s}$$

The speed of an accelerating object can be determined by using a tangent to measure the gradient of the distance–time graph.

Acceleration

Acceleration can be calculated using the following equation:

$$\text{average acceleration} = \frac{\text{change in velocity}}{\text{time taken}}$$

$$\left[a = \frac{\Delta v}{t} \right]$$

➤ acceleration, a, in metres per second squared, m/s²
➤ change in velocity, Δv, in metres per second, m/s
➤ time, t, in seconds, s

Example:
A car travels from 0 to 27 m/s in 3.8 seconds. What is its acceleration?

$$\text{change in velocity} = 27 - 0 = 27 \text{ m/s}$$

$$\text{average acceleration} = \frac{\text{change in velocity}}{\text{time taken}}$$

$$= \frac{27}{3.8} = 7.1 \text{ m/s}^2$$

An object that slows down (decelerates) has a negative acceleration.

Measure out a short distance and roll a ball. Time how long it takes the ball to travel the distance you've measured. Plot the distance travelled and the time taken on a graph. Repeat the investigation, but this time roll the ball harder. Again, plot the result on the same speed–time graph. Compare the gradients of the two lines. What do you notice? Use your graph to calculate the average speed of the two balls.

Velocity–time graph

Acceleration can be calculated from the gradient of a velocity–time graph.

HT Distance travelled can be calculated from the area under a velocity–time graph.

The graph below shows the movement of a car.

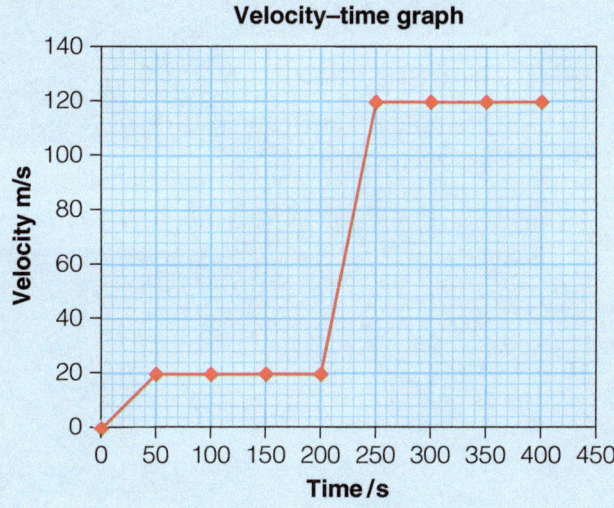

Velocity–time graph

The acceleration between 0–50 seconds can be worked out as follows:

$$\text{gradient of line} = \frac{\text{velocity}}{\text{time}}$$

$$= \frac{20}{50} = 0.4 \text{ m/s}^2$$

As the gradient of the line is steeper between 200–250 seconds than it is between 0–50 seconds, the acceleration must have been greater between 200–250 seconds.

HT The distance travelled between 0–200 seconds can be worked out as follows:

Area from 0–50s

$$= \left(20 \times \frac{50}{2}\right) = 500 \text{ m}$$

Area from 50–200s

$$= 20 \times 150 = 3000 \text{ m}$$

Total distance travelled $= 3500$ m

The equation below applies to uniform motion:

final velocity^2 – initial velocity^2 = 2 × acceleration × distance

$$v^2 - u^2 = 2\,as$$

➤ final velocity, v, in metres per second, m/s
➤ initial velocity, u, in metres per second, m/s
➤ acceleration, a, in metres per second squared, m/s^2
➤ distance, s, in metres, m

Falling objects

Near the Earth's surface, any object falling freely under gravity has an acceleration of about 10 m/s^2. This is the gravitational field strength, or acceleration due to gravity, and is used to calculate weight.

An object falling through a fluid initially accelerates due to the force of gravity. Eventually the resultant force will be zero and the object will move at its terminal velocity.

1. What does the gradient of a distance–time graph represent?
2. What type of acceleration will an object which is slowing down have?
HT 3. What does the area under a velocity–time graph represent?

Keyword

Acceleration ➤ Change in velocity over time

Mind map

Contact vs non-contact

Spring constant

Scalar vs vector

Work done and energy transfer

Elastic deformation

Forces

Elasticity

FORCES

Resultant forces

Inelastic deformation

Displacement

Pressure in a fluid

Velocity

Distance

Upthrust

Atmospheric pressure

Acceleration

Speed

Practice questions

1. a) The gravitational field strength on Mars is 3.7 N/kg.

    What would be the weight of a Mars rover with a mass of 187 kg?   **(2 marks)**

    b) Would the weight of the rover be different on Earth?
    Explain your answer.   **(2 marks)**

    c) A water container on Mars has a height of 3 m. Water has a density of 1000 kg/m³.

    What is the pressure at the bottom of the container?   **(2 marks)**

2. The graph below shows the velocity of an object.

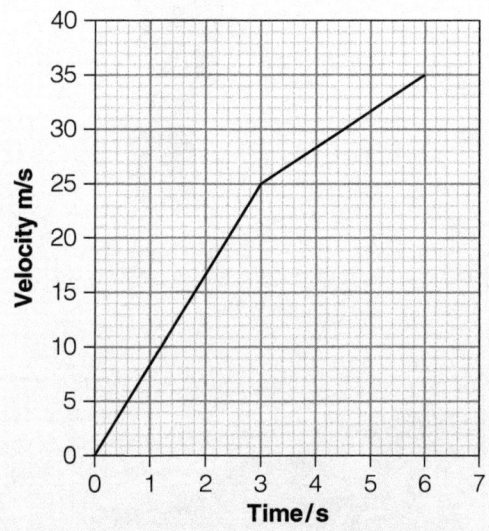

    a) Use the graph to calculate the:

        i)   acceleration from 0–3 seconds.   **(2 marks)**

        ii) distance travelled in 3 seconds.   **(2 marks)**

    b) Was the acceleration between 0–3 seconds greater than the acceleration
    between 3–6 seconds?
    Explain your answer.   **(2 marks)**

Newton's laws

Newton's first law

Newton's first law deals with the effect of resultant forces.

➤ If the resultant force acting on an object is zero and the object is stationary, the object remains stationary.

➤ If the resultant force acting on an object is zero and the object is moving, the object continues to move at the same speed and in the same direction (its velocity will stay the same).

Newton's first law means that the velocity of an object will only change if a resultant force is acting on the object.

Examples of resultant force:

Stationary object – zero resultant force

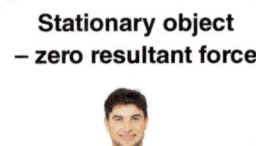

Object moving at constant speed – zero resultant force

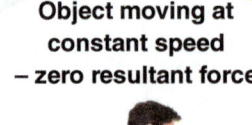

A car is stationary. At this point the resultant force is zero.

The car starts to move and accelerates. As it is accelerating, the resultant force on the car is no longer zero.

The car travels at a constant velocity. The resultant force is zero again.

HT The tendency of objects to continue in their state of rest or of uniform motion is called **inertia**.

Newton's second law

The acceleration of an object is proportional to the resultant force acting on the object, and inversely proportional to the mass of the object.

Therefore:

> **acceleration** ∝ **resultant force**
> **resultant force = mass × acceleration**
> ➤ force, *F*, in newtons, N
> ➤ mass, *m*, in kilograms, kg
> ➤ acceleration, *a*, in metres per second squared, m/s²

Example:
A motorbike and rider of mass 270 kg accelerate at 6.7 m/s². What is the resultant force on the motorbike?

$$270 \times 6.7 = 1809 \text{ N}$$

If the motorbike slows to a constant speed, the resultant force would now be 0 (as acceleration = 0, $270 \times 0 = 0$).

HT **Inertial mass** is a measure of how difficult it is to change the velocity of an object. It is defined by the ratio of force over acceleration.

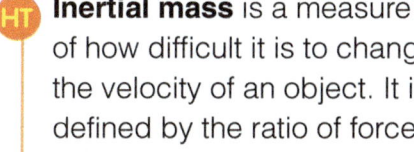
Make a revision poster with a drawing illustrating each of Newton's laws and the resultant force equation.

Keyword

HT **Inertia** ➤ The tendency of objects to continue in their state of rest or uniform motion

Newton's third law

Whenever two objects interact, the forces they exert on each other are equal and opposite.

When a fish swims it exerts a force on the water, pushing it backwards. The water exerts an equal and opposite force on the fish, pushing it forwards.

1. A book of weight 67 N is on a desk. What force is the desk exerting on the book?
2. A sprinter of mass 89 kg accelerates at 10 m/s². What is the resultant force?
3. What happens to the speed of a moving object that has a resultant force of 0 acting on it?

Forces and braking

Stopping distance

The stopping distance of a vehicle is a sum of the **thinking distance** and the **braking distance**.

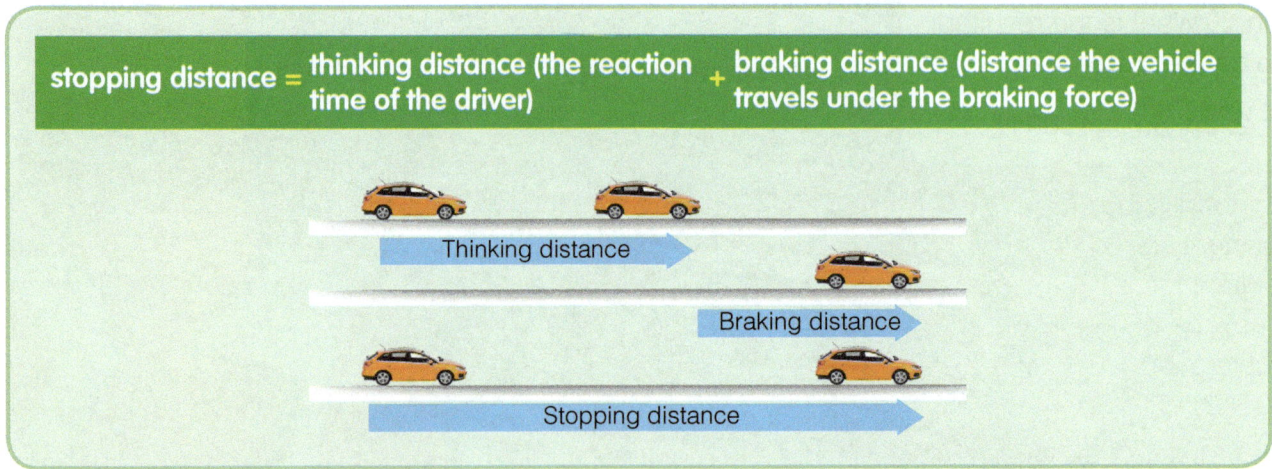

| stopping distance = | thinking distance (the reaction time of the driver) | + | braking distance (distance the vehicle travels under the braking force) |

A greater vehicle speed leads to a greater stopping distance (given a set braking force). As the vehicle is going faster, the car will travel further in the time taken for the driver to react and apply the brakes.

The graph below shows the stopping distances over a range of speeds for a car.

Typical stopping distances

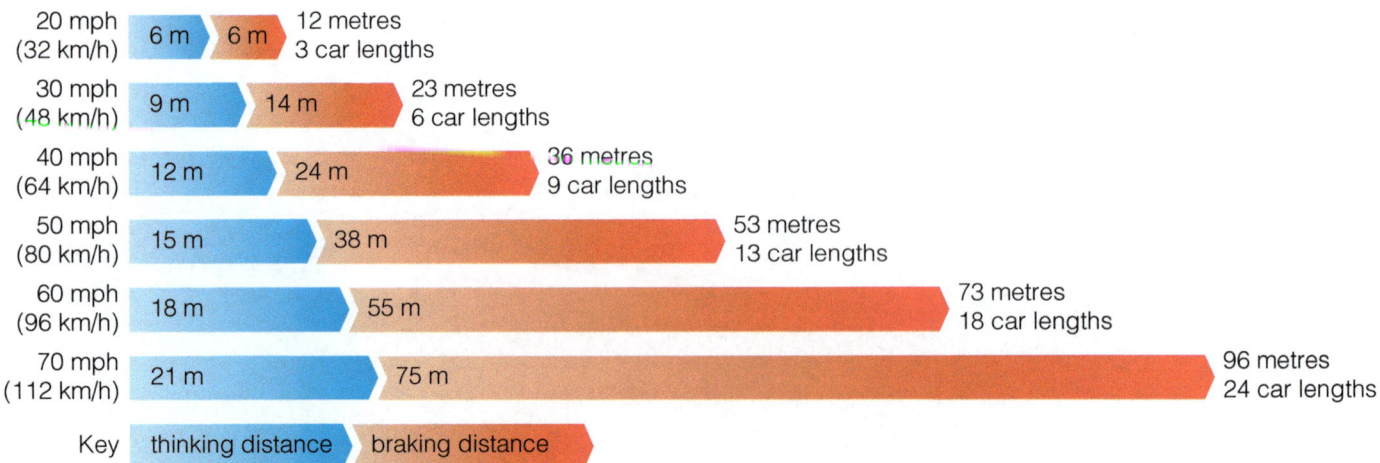

Speed	Thinking distance	Braking distance	Total
20 mph (32 km/h)	6 m	6 m	12 metres / 3 car lengths
30 mph (48 km/h)	9 m	14 m	23 metres / 6 car lengths
40 mph (64 km/h)	12 m	24 m	36 metres / 9 car lengths
50 mph (80 km/h)	15 m	38 m	53 metres / 13 car lengths
60 mph (96 km/h)	18 m	55 m	73 metres / 18 car lengths
70 mph (112 km/h)	21 m	75 m	96 metres / 24 car lengths

Key: thinking distance   braking distance

Braking distance

The braking distance of a vehicle can be affected by adverse conditions outside and a poor vehicle condition.

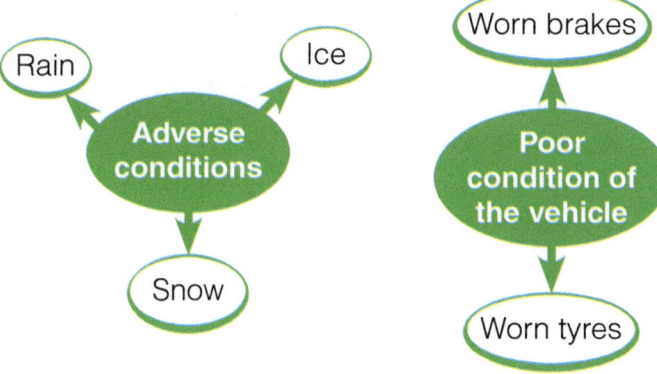

Reaction time

The average reaction time is between 0.4–0.9 seconds but this varies from person to person. Thinking distance increases as speed increases. This is because a faster vehicle travels further during the time taken to react.

A driver's reaction time can be affected by tiredness, drugs and alcohol. Distractions, such as eating, drinking or using a mobile phone, may also affect a driver's ability to react.

Design a road safety poster summarising stopping distance, thinking distance, braking distance and all the factors that affect them.

Keywords

Thinking distance ➤ Distance travelled before the driver applies the brakes
Braking distance ➤ Distance a vehicle travels after the brakes have been applied

Braking force

To slow a vehicle down, the friction between the tyres and the road must be increased. This is done by the brakes.

This is what happens when a force is applied to the brakes of a vehicle.

Work is done by the friction force between the brakes and the wheel.	This reduces the kinetic energy of the vehicle.	The temperature of the brakes increases.

➤ The greater the speed of a vehicle, the greater the braking force needed to stop the vehicle in a certain distance.
➤ The greater the braking force, the greater the deceleration of the vehicle.
➤ The work done to stop the vehicle is equal to the initial kinetic energy of the vehicle.

Large decelerations may lead to brakes overheating and/or loss of control.

1. What two distances make up the stopping distance?
2. Give two examples of adverse road conditions which can increase the stopping distance.
3. Why is a large deceleration potentially dangerous?

Momentum

Momentum

Moving objects have **momentum**. Momentum is given by the following equation:

> momentum = mass × velocity
>
> $p = mv$
>
> ➤ momentum, p, in kilograms metre per second, kg m/s
> ➤ mass, m, in kilograms, kg
> ➤ velocity, v, in metres per second, m/s

Example:
A car of mass 1100 kg is travelling at 13 m/s. What is its momentum?

$p = mv$
$ = 1100 \times 13$
$ = 14\,300$ kg m/s

Conservation of momentum

In a closed system, the total momentum before an event is equal to the total momentum after the event. This is called **conservation of momentum**.

Example:
A cue ball with mass of 0.17 kg travelling at 4 m/s hits a stationary snooker ball with a mass of 0.16 kg. The first ball stops while the other ball starts to move. What is the velocity of the second ball?

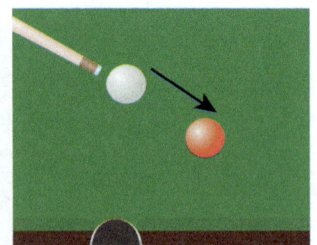
Before collision

After collision

momentum = mass × velocity

momentum of first ball before collision $= 0.17 \times 4 = 0.68$ kg m/s

momentum of first ball = 0 (as it is now stationary) As momentum is the same before and after the collision, the second ball has a momentum of 0.68 kg m/s.

$0.68 = \text{mass} \times \text{velocity}$
$0.68 = 0.16 \times \text{velocity}$
$\dfrac{0.68}{0.16} = \text{velocity}$
$\text{velocity} = 4.25$ m/s

Changes in momentum

When a force acts on an object that is moving or is able to move, a change in momentum occurs.

The equations $F = m \times a$ and

$$a = \frac{v - u}{t}$$

lead to the equation

$$F = \frac{m\Delta v}{\Delta t}$$

➤ $m\Delta v$ = the rate of change of momentum

Example:

When a 20 g dart travelling at 18 m/s hits a dartboard it comes to rest in 0.2 s. What force is generated by this impact?

initial momentum of dart = mass × velocity
$$= 18 \times 0.02$$
$$= 0.36 \text{ kg m/s}$$

final momentum = 0 kg m/s
change in momentum = 0.36 − 0
$$= 0.36$$

force $= \dfrac{\text{the rate of change of momentum}}{\text{time}}$

$$= \frac{0.36}{0.2}$$

$$= 1.8 \text{ N}$$

Safety features

Safety features include:

➤ airbags
➤ seat belts
➤ gymnasium crash mats
➤ cycle helmets
➤ cushioned surfaces for playgrounds.

All these features slow down the change in momentum in a collision. This reduces the forces on the people involved, reducing injury.

Keyword

Momentum ➤ Product of an object's mass and velocity

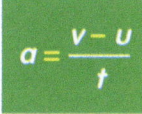

Try catching a table tennis ball and then a cricket ball. Note how much harder it is to catch the heavier cricket ball. Use force, mass and momentum to explain this difference.

1. How do airbags reduce the chance of injury in a collision?
2. What is the momentum of an object of 6 kg moving at 4 m/s?
3. What happens to the momentum when a force acts on a moving object?

Mind map

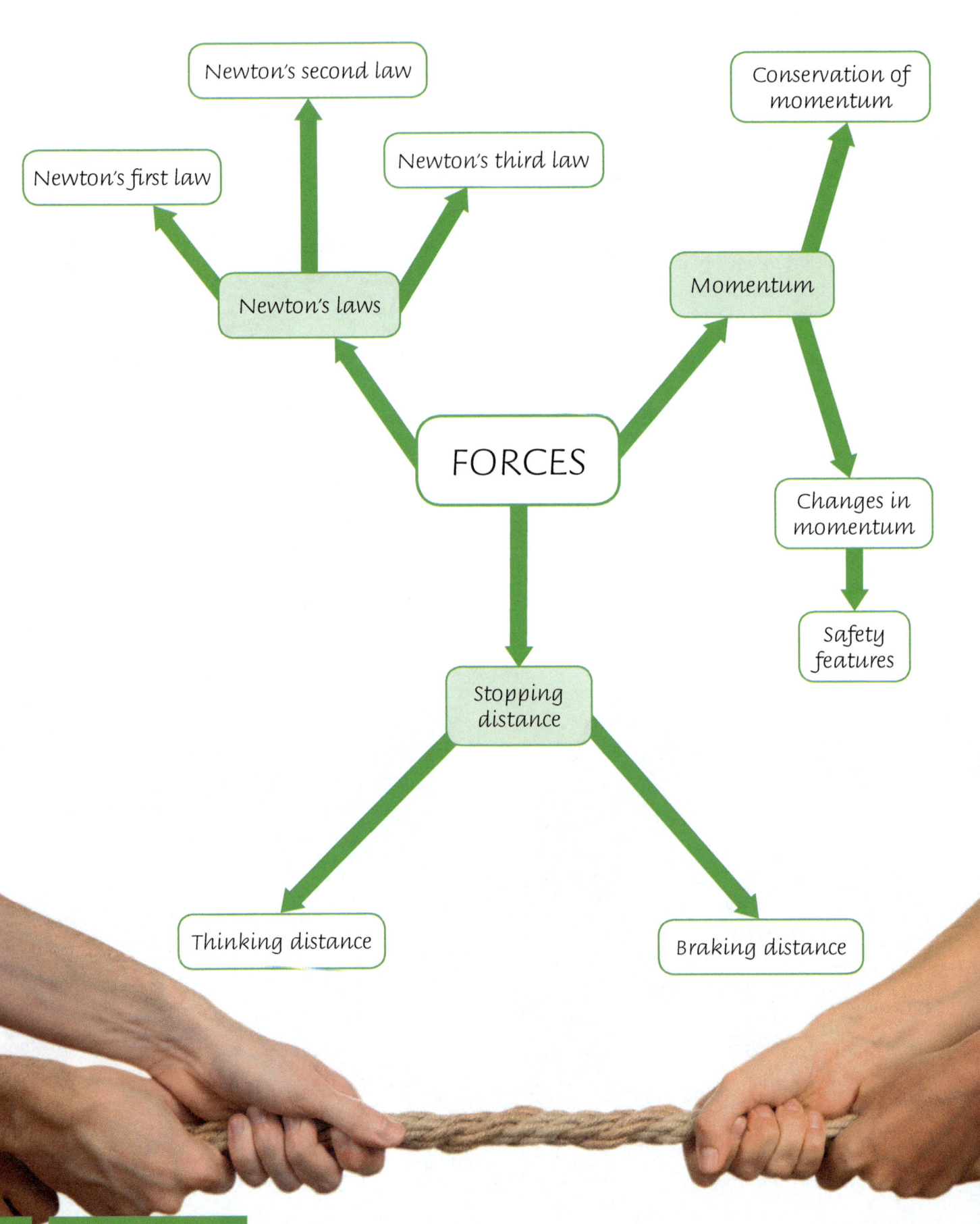

Newton's second law

Newton's first law

Newton's third law

Newton's laws

Conservation of momentum

Momentum

FORCES

Changes in momentum

Safety features

Stopping distance

Thinking distance

Braking distance

Practice questions

1. **a)** The diagram below shows the forces acting on a stationary bike and rider.

   What is the upward force from the ground? Explain how you arrived at your answer. **(2 marks)**

890 N

   b) The diagram on the right shows the same bike and rider moving.

   Is the rider accelerating, decelerating or moving at the same speed? Explain, using Newton's laws, how you arrived at your answer. **(2 marks)**

25 N   25 N

   c) A change in terrain caused the friction to increase. What effect would this have on the resultant force and the speed of the rider? **(2 marks)**

2. A car in a crash test has a total mass of 1200 kg. The car is travelling at a speed of 16 m/s.

   a) The car accelerates at 2.5 m/s^2. What is its resultant force? **(2 marks)**

   b) The car maintains a constant speed of 16 m/s. What is the total momentum of the car? **(3 marks)**

   c) When the car crashes, its momentum becomes zero in 0.4 seconds. What is the force generated by this crash? **(3 marks)**

   d) Give two safety features that could reduce the risk of injury in a collision like this. **(2 marks)**

Changes in energy

Keyword

Specific heat capacity ➤ Amount of energy required to raise the temperature of one kilogram of a substance by one degree Celsius

A system is an object or group of objects. When a system changes, so does the energy stored within its objects.

an object projected upwards, e.g a rocket – chemical energy in the fuel is converted to kinetic energy and gravitational potential energy

bringing water to a boil in an electric kettle – electrical energy to heat and sound energy

Changes

a moving object hitting an obstacle – kinetic energy to sound, heat and kinetic energy

an object accelerated by a constant force, e.g a tennis player hitting a tennis ball – chemical energy in the player to kinetic energy in their arm to kinetic energy in the racket and then kinetic energy in the ball

a vehicle slowing down – kinetic energy to sound and heat energy (friction)

Energy in moving objects

The kinetic energy of a moving object can be calculated using the following equation:

kinetic energy = 0.5 × mass × (speed)^2

$$E_k = \frac{1}{2}mv^2$$

➤ kinetic energy, E_k, in joules, J
➤ mass, m, in kilograms, kg
➤ speed, v, in metres per second, m/s

Example:
A ball of mass 0.3 kg falls at 10 m/s. What is the ball's kinetic energy?

$$E_k = \frac{1}{2}mv^2$$

$$= 0.5 \times 0.3 \times 10^2$$

$$= 15\ J$$

Create a large formula poster with each of the formulas on these pages, a worked example different to the ones on this page and a diagram showing the energy transfer.

Elastic potential energy (E_e)

$$\text{elastic potential energy} = 0.5 \times \text{spring constant} \times \text{extension}^2$$

$$E_e = \frac{1}{2}ke^2$$

(assuming the limit of proportionality has not been exceeded)

➤ elastic potential energy, E_e, in joules, J
➤ spring constant, k, in newtons per metre, N/m
➤ extension, e, in metres, m

Example:

A spring has a spring constant of 500 N/m and is extended by 0.2 m. What is the elastic potential energy in the spring?

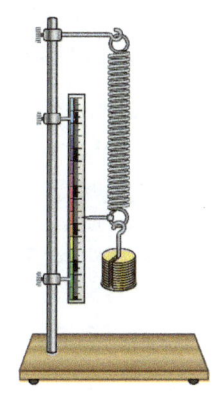

$$E_e = \frac{1}{2}ke^2$$
$$= \frac{1}{2}500 \times 0.2^2$$
$$= 250 \times 0.04$$
$$= 10 \text{ J}$$

Gravitational potential energy ($g.p.e.$)

The amount of gravitational potential energy gained by an object raised above ground level can be calculated using the following equation:

$$g.p.e. = \text{mass} \times \text{gravitational field strength} \times \text{height}$$

$$E_p = mgh$$

➤ gravitational potential energy, E_p, in joules, J
➤ mass, m, in kilograms, kg
➤ gravitational field strength, g, in newtons per kilogram, N/kg
➤ height, h, in metres, m

Example:

A car of mass 1500 kg drives up a hill which is 67 m high. What is the gravitational potential energy of the car at the top of the hill? (Assume the gravitational field strength is 10 N/kg)

$$E_p = mgh$$
$$= 15\,000 \times 67 \times 10$$
$$= 10\,050 \text{ kJ}$$

Changes in thermal energy

The **specific heat capacity** of a substance is the amount of energy required to raise the temperature of one kilogram of the substance by one degree Celsius. The specific heat capacity can be used to calculate the amount of energy stored in or released from a system as its temperature changes. This can be calculated using the following equation:

$$\text{change in thermal energy} = \text{mass} \times \text{specific heat capacity} \times \text{temperature change}$$

$$\Delta E = mc\,\Delta\theta$$

➤ change in thermal energy, ΔE, in joules, J
➤ mass, m, in kilograms, kg
➤ specific heat capacity, c, in joules per kilogram per degree Celsius, J/kg °C
➤ temperature change, $\Delta\theta$, in degrees Celsius, °C

Example:

0.75 kg of 100°C water cools to 23°C. What is the change in thermal energy?
(Specific heat capacity of water = 4184 J/kg °C)

$$\Delta E = mc\,\Delta\theta$$
$$= 0.75 \times 4184 \times (100 - 23)$$
$$= 241\,626 \text{ J} = 242 \text{ kJ}$$

1. What is the kinetic energy of a 65 g object travelling at 6 m/s?
2. What is the gravitational potential energy of a 17 kg object at a height of 987 m? (gravitational field strength = 10 g)
3. What is specific heat capacity?

Conservation and dissipation of energy

Keyword

Thermal insulation ➤ Material used to reduce transfer of heat energy

Energy transfers in a system

Energy can be **transferred**, **stored** or **dissipated**. It cannot be created or destroyed.

In a closed system there is no net change to the total energy when energy is transferred.

Only part of the energy is usefully transferred. The rest of the energy dissipates and is transferred in less useful ways, often as heat or sound energy. This energy is considered **wasted**.

Examples of processes which cause a rise in temperature and so waste energy as heat include:

➤ friction between the moving parts of a machine
➤ electrical work against the resistance of connecting wires.

If the energy that is wasted can be reduced, that means more energy can be usefully transferred. The less energy wasted, the more efficient the transfer.

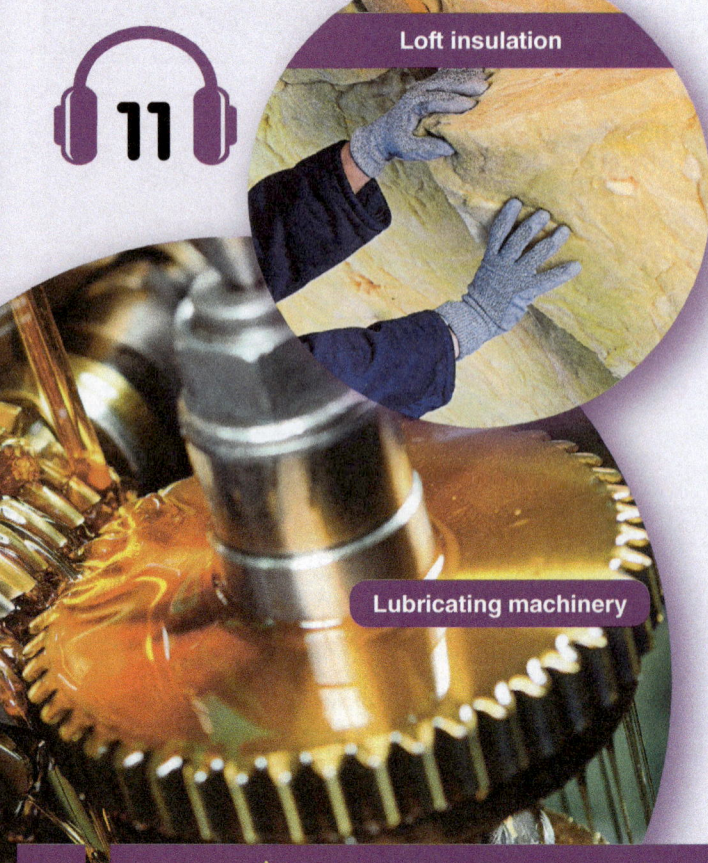

Loft insulation

Lubricating machinery

Efficiency

The energy efficiency for any energy transfer can be calculated using the following equation:

$$efficiency = \frac{useful\ output\ energy\ transfer}{useful\ input\ energy\ transfer}$$

Efficiency may also be calculated using the following equation:

$$efficiency = \frac{useful\ power}{total\ power\ output}$$

Example:

Kat uses a hairdryer. Some of the energy is wasted as sound. The electrical energy input is 24 kJ. The energy wasted is 7 kJ. What is the efficiency of the hairdryer?

First calculate the useful energy transferred.

useful energy = total energy – wasted energy
= 24 – 7 = 17 kJ

Now calculate the efficiency.

$$efficiency = \frac{useful\ output\ energy\ transfer}{useful\ input\ energy\ transfer}$$
$$= \frac{17}{24} = 0.71$$

This decimal efficiency can be represented as a percentage by multiplying it by 100.

$$0.71 \times 100 = 71\%$$

There are many ways energy efficiency can be increased:

➤ Lubrication, thermal insulation and low resistance wires reduce energy waste and improve efficiency.
➤ Thermal insulation, such as loft insulation, reduces heat loss.
➤ Low resistance wires reduce energy lost as heat when an electrical current flows through them.

Power

One of the definitions of power is work done over time. The power equation is:

$$power = \frac{work\ done}{time} \qquad P = \frac{W}{T}$$

- ➤ P = power (watts)
- ➤ E = work done (joules)
- ➤ T = time (seconds)

Example:

A weightlifter is lifting weights that have a mass of 80 kg. What power is required to lift them 2 metres vertically in 4 seconds? (Assume gravitational field strength of 10 N/kg)

$$power = \frac{work\ done}{time}$$

work done = force × distance

force = mass × gravitational field strength

= 80 × 10 = 800 N

work done = 800 × 2 = 1600 J

$$power = \frac{1600}{4}$$

= 800 W

A second weightlifter lifted the same mass to the same height in 3 seconds. As he carried out the same amount of work but in a shorter time, he would have a greater power.

Thermal insulation

Thermal insulation has a low thermal conductivity, so has a slow rate of energy transfer by conduction. U-values give a measure of the heat loss through a substance. A higher U-value indicates that a material has a higher thermal conductivity.

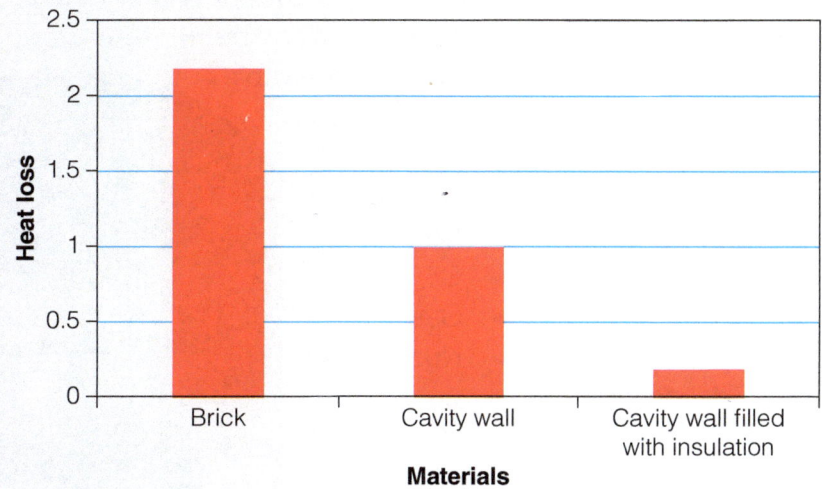

Changing the material of walls to materials that have a lower thermal conductivity reduces heat loss (the U-value) and so a building cools more slowly, reducing heating costs. The graph below shows the heat lost by different types of wall.

Heat loss (y-axis, scale 0 to 2.5)

Materials (x-axis): Brick, Cavity wall, Cavity wall filled with insulation

The house below is not fitted with insulation, so lots of heat is being lost through the walls due to their high thermal conductivity. This can be seen by the red/orange colour on the infrared image.

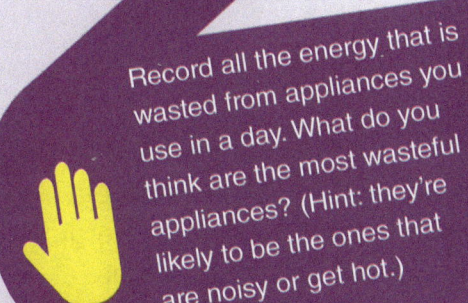

Record all the energy that is wasted from appliances you use in a day. What do you think are the most wasteful appliances? (Hint: they're likely to be the ones that are noisy or get hot.)

1. Why is it important to increase the efficiency of an energy transfer?
2. A washing machine transfers 400 J of useful energy out of a total of 652 J. What is the efficiency of the washing machine?
3. What effect does fitting cavity wall insulation have on the thermal conductivity of the building's wall?

National and global energy resources

Keywords
Renewable ➤ Energy resource that can be replenished as it is used
Non-renewable ➤ Energy resource that will eventually run out

Energy resources

Energy resources are used for transport, electricity generation and heating.

Energy resources can be divided into **renewable** and **non-renewable**. Renewable resources can be replenished as they are used. Non-renewable energy resources will eventually run out.

Renewable	Non-renewable
➤ Bio-fuel	➤ Coal
➤ Wind	➤ Oil
➤ Hydro-electricity	➤ Gas
➤ Geothermal	➤ Nuclear fuel
➤ Tidal power	
➤ Solar power	
➤ Water waves	

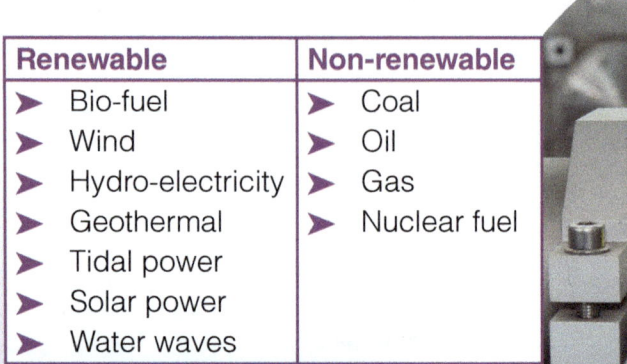

Non-renewable

Renewable

12

Reliability

Fossil fuels (coal, oil and gas) and nuclear fuel are very reliable as they can always be used to release energy. Fossil fuels are burnt to release the stored chemical energy, and nuclear fuel radiates energy.

The wind and the Sun are examples of energy resources that are not very reliable, as the wind doesn't always blow and it's not always sunny.

All the energy resources in the table on the opposite page can be used to generate electricity. The following energy resources can be used for transport and heating.

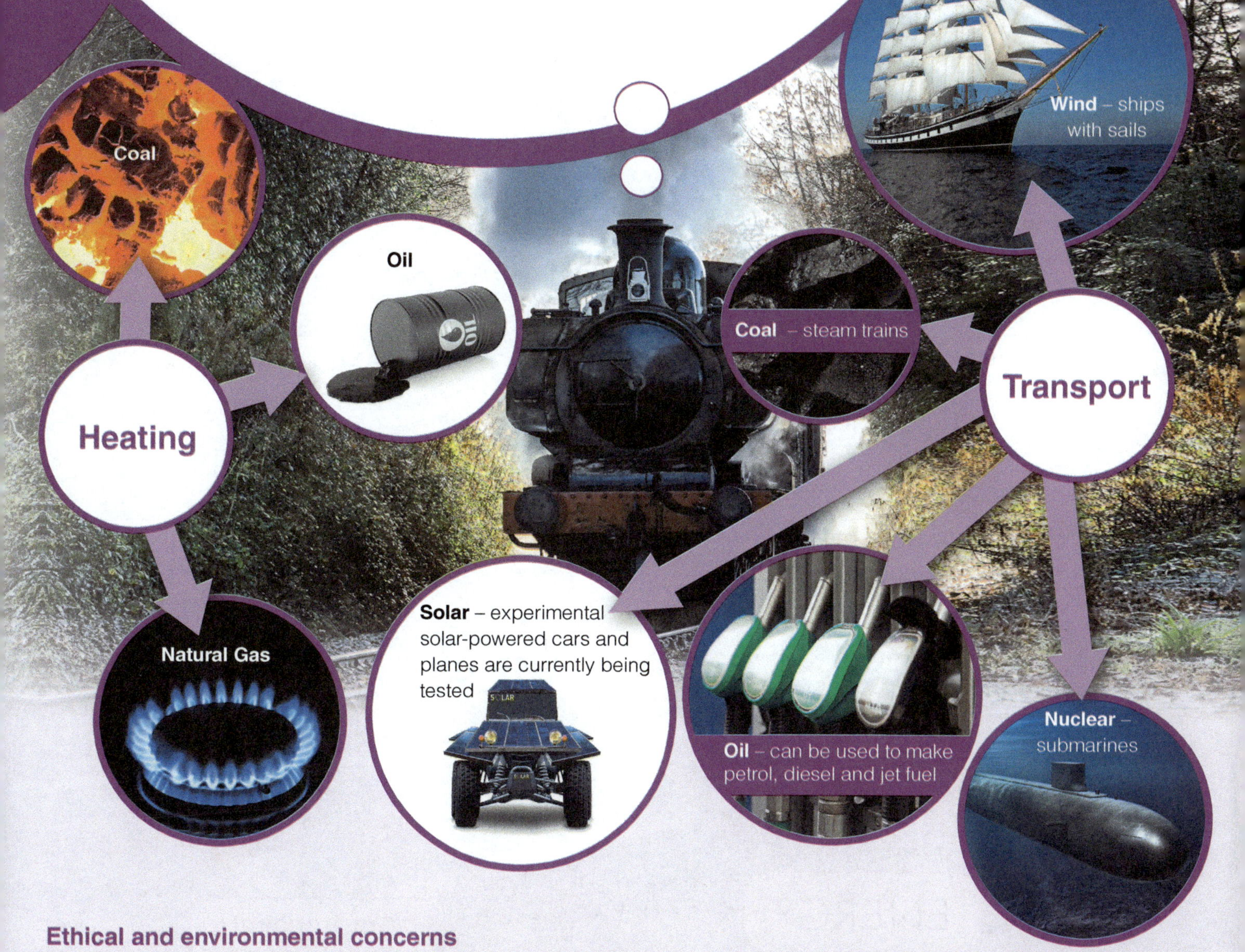

Wind – ships with sails

Coal

Oil

Coal – steam trains

Heating

Transport

Natural Gas

Solar – experimental solar-powered cars and planes are currently being tested

Oil – can be used to make petrol, diesel and jet fuel

Nuclear – submarines

Ethical and environmental concerns

➤ Burning fossil fuels produces carbon dioxide, which is a greenhouse gas. Increased greenhouse gas emissions are leading to climate change. Particulates and other pollutants are also released, which cause respiratory problems.

➤ Nuclear power produces hazardous nuclear waste and also has the potential for nuclear accidents, with devastating health and environmental consequences.

➤ Some people consider wind turbines to be ugly and to spoil the landscape.

➤ Building tidal power stations can lead to the destruction of important tidal habitats.

In recent years, there have been attempts to move away from over-reliance on fossil fuels to renewable forms of energy. However, fossil fuels are still used to deliver the vast majority of our energy needs.

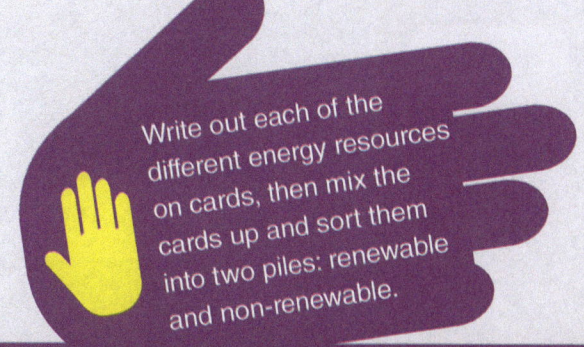

Write out each of the different energy resources on cards, then mix the cards up and sort them into two piles: renewable and non-renewable.

1. Give three examples of renewable energy resources and three examples of non-renewable energy resources.
2. Why aren't solar and wind power reliable energy resources?
3. What is an environmental concern about burning fossil fuels?

Mind map

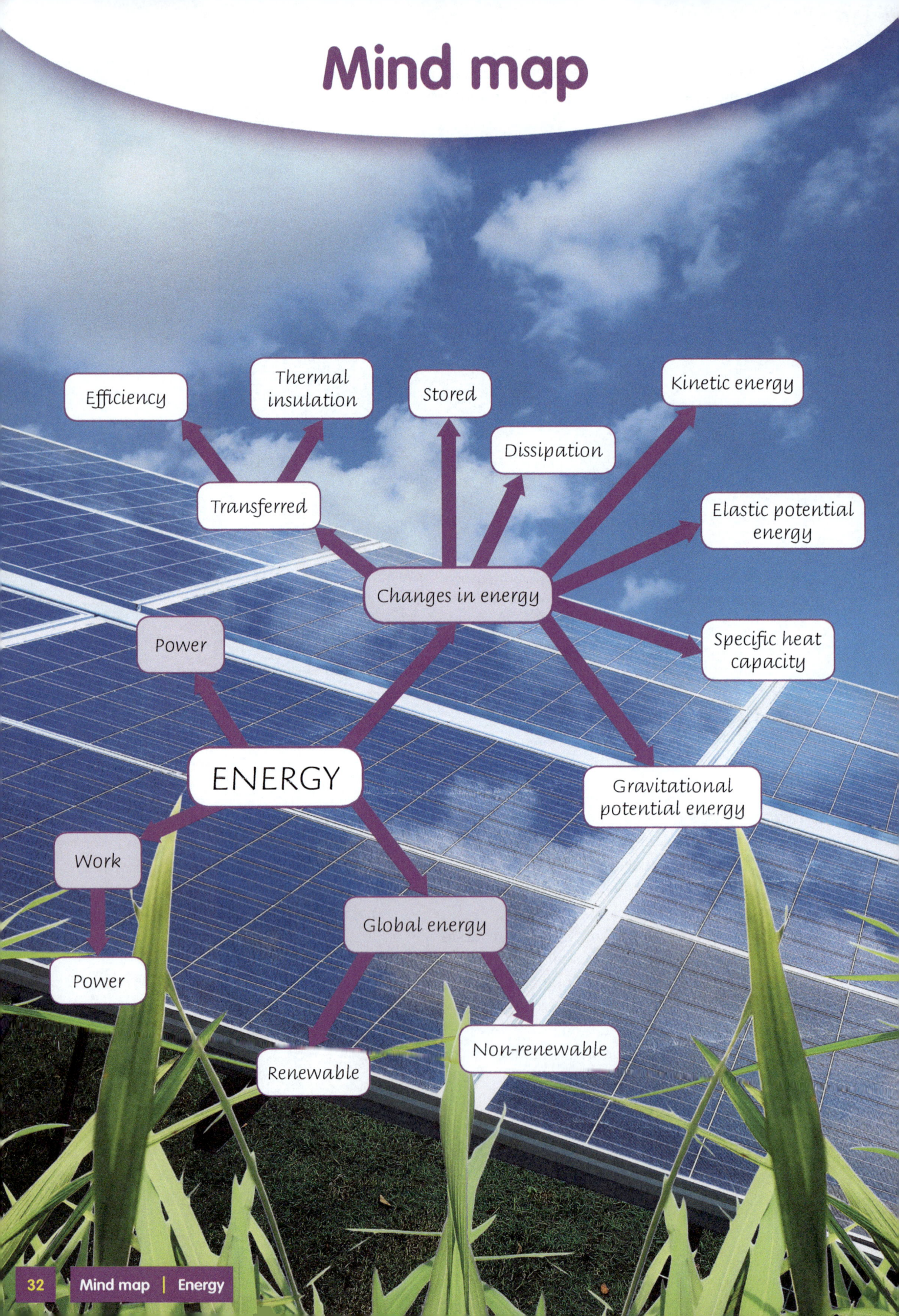

Efficiency

Thermal insulation

Stored

Kinetic energy

Dissipation

Transferred

Elastic potential energy

Changes in energy

Power

Specific heat capacity

ENERGY

Gravitational potential energy

Work

Power

Global energy

Renewable

Non-renewable

Practice questions

1. **a)** Explain the difference between a renewable and a non-renewable energy resource. **(2 marks)**

    b) Nuclear power is a very reliable source of energy but some environmentalists protest against it.
    Explain why. **(2 marks)**

    c) Many environmentalists would rather use wind or tidal power stations.
    What are the potential disadvantages of using these energy resources? **(2 marks)**

2. A car has a mass of 1600 kg and moves at 32 m/s.

    a) What is the kinetic energy of the car? **(2 marks)**

    b) If the fuel burnt by the car to reach this speed contained 1500 kJ, what was the efficiency of this transfer of energy? **(2 marks)**

    c) Give two ways energy was transferred into less useful forms by the car's engine. **(2 marks)**

    d) Give one feature of the car's engine that reduced the energy dissipated. **(1 mark)**

3. A bungee jumper of mass 77 kg stands on a bridge 150 m high.

    a) What is their gravitational potential energy? (Assume gravitational field strength = 10 N/kg) **(2 marks)**

    b) The bungee jumper jumps off and quickly reaches a speed of 20 m/s.
    What is their kinetic energy at this point? **(2 marks)**

    c) Explain the energy transfers that occur from when the bungee jumper jumps to when they reach their lowest height. **(2 marks)**

Transverse and longitudinal waves

13

Transverse waves

In a transverse wave, the oscillations are perpendicular to the direction of energy transfer, such as the ripples on the surface of water.

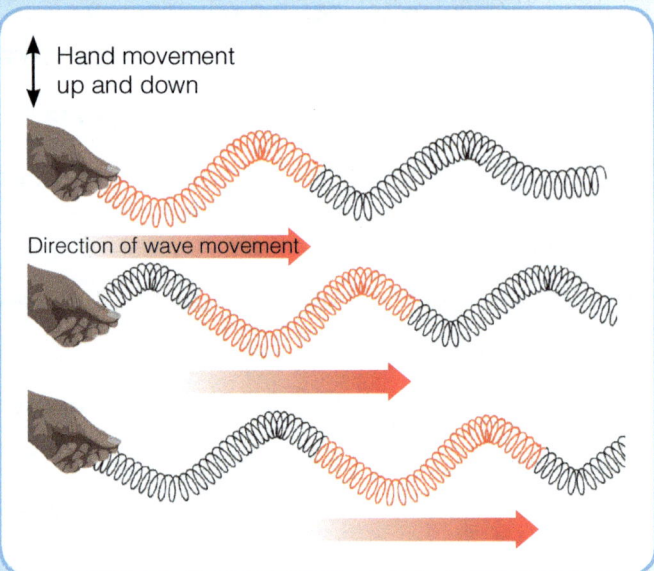

Hand movement up and down

Direction of wave movement

Longitudinal waves

In a longitudinal wave, the oscillations are parallel to the direction of energy transfer. Longitudinal waves show areas of **compression** and **rarefaction**, such as sound waves travelling through air.

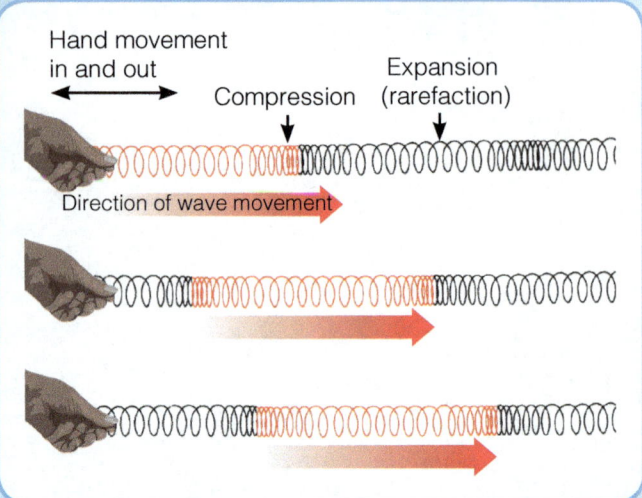

Hand movement in and out

Compression

Expansion (rarefaction)

Direction of wave movement

Wave movement

In both sound waves in air and ripples on the water surface, it is the wave that moves forward rather than the air or water molecules. The waves transfer energy and information without transferring matter.

This can be shown experimentally. For example, when a tuning fork is used to create a sound wave that moves out from the fork, the air particles don't move away from the fork. (This would create a vacuum around the tuning fork.)

Properties of waves

Waves are described by their:

➤ **amplitude** – The amplitude of a wave is the maximum displacement of a point on a wave away from its undisturbed position.

➤ **wavelength** – The wavelength of a wave is the distance from a point on one wave to the equivalent point on the adjacent wave.

➤ **frequency** – The frequency of a wave is the number of waves passing a point each second.

➤ **period** – The time for one complete wave to pass a fixed point. The equation for the time period (T) of a wave is given by the following equation:

$$\text{period} = \frac{1}{\text{frequency}}$$

$$\text{period} = \frac{1}{f}$$

➤ period, T, in seconds, s

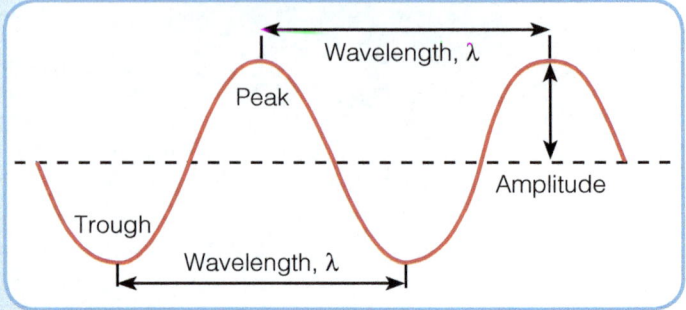

Wavelength, λ

Peak

Trough

Amplitude

Wavelength, λ

Keywords

Compression ➤ A region in a longitudinal wave where the particles are closer together

Rarefaction ➤ A region in a longitudinal wave where the particles are further apart

Wave speed

The wave speed or wave velocity is the speed at which the energy is transferred (or the wave moves) through the medium.

The wave speed is given by the following wave equation:

wave speed = frequency × wavelength

$$v = f\lambda$$

➤ wave speed, v, in metres per second, m/s
➤ frequency, f, in hertz, Hz
➤ wavelength, λ, in metres, m

Example:

A sound wave in air has a frequency of 250 Hz and a wavelength of 1.32 m. What is the speed of the sound wave?

wave speed = frequency × wavelength
= 250 × 1.32
= 330 m/s

A ripple tank and a stroboscope can be used to measure the speed of ripples on the surface of water.

As wave speed, frequency and wavelength are all interrelated, changes in wave speed due to waves travelling between different media will affect the wavelength of a sound. For example, when sound waves enter a more dense medium, the wave speed increases. This leads to an increase in the sound's wavelength.

Make a series of cards with the names of key properties of a wave on one side and their definitions on the other. Use these cards for revision and to test yourself.

1. Give an example of a longitudinal wave and an example of a transverse wave.
2. Define the term 'wavelength'.
3. A sound wave travels in air at a speed of 343 m/s at a frequency of 90 Hz. The sound wave enters seawater where its speed increases to 1500 m/s. If the frequency of the water remains constant, what is the effect on the wavelength?

When a sound wave reaches a boundary between two different materials, the waves can be **absorbed** (energy transferred to the matter), **transmitted** (passed through the material) or **reflected**.

Sound waves and reflection of waves

14

Reflection

The ray diagram right shows the **reflection** of a wave from a surface.

➤ The incident ray is the incoming ray and the refracted ray is created by the reflection.
➤ The normal is a line perpendicular to the reflecting surface at the point where the incident ray hits the surface.
➤ The **angle of incidence** is the angle between the incident ray and the normal.
➤ The **angle of reflection** is the angle between the reflected ray and the normal.
➤ The angle of incidence = the angle of reflection.
➤ If the angle of incidence of the wave is above the critical angle, all the waves reflect back. This is **total internal reflection**.

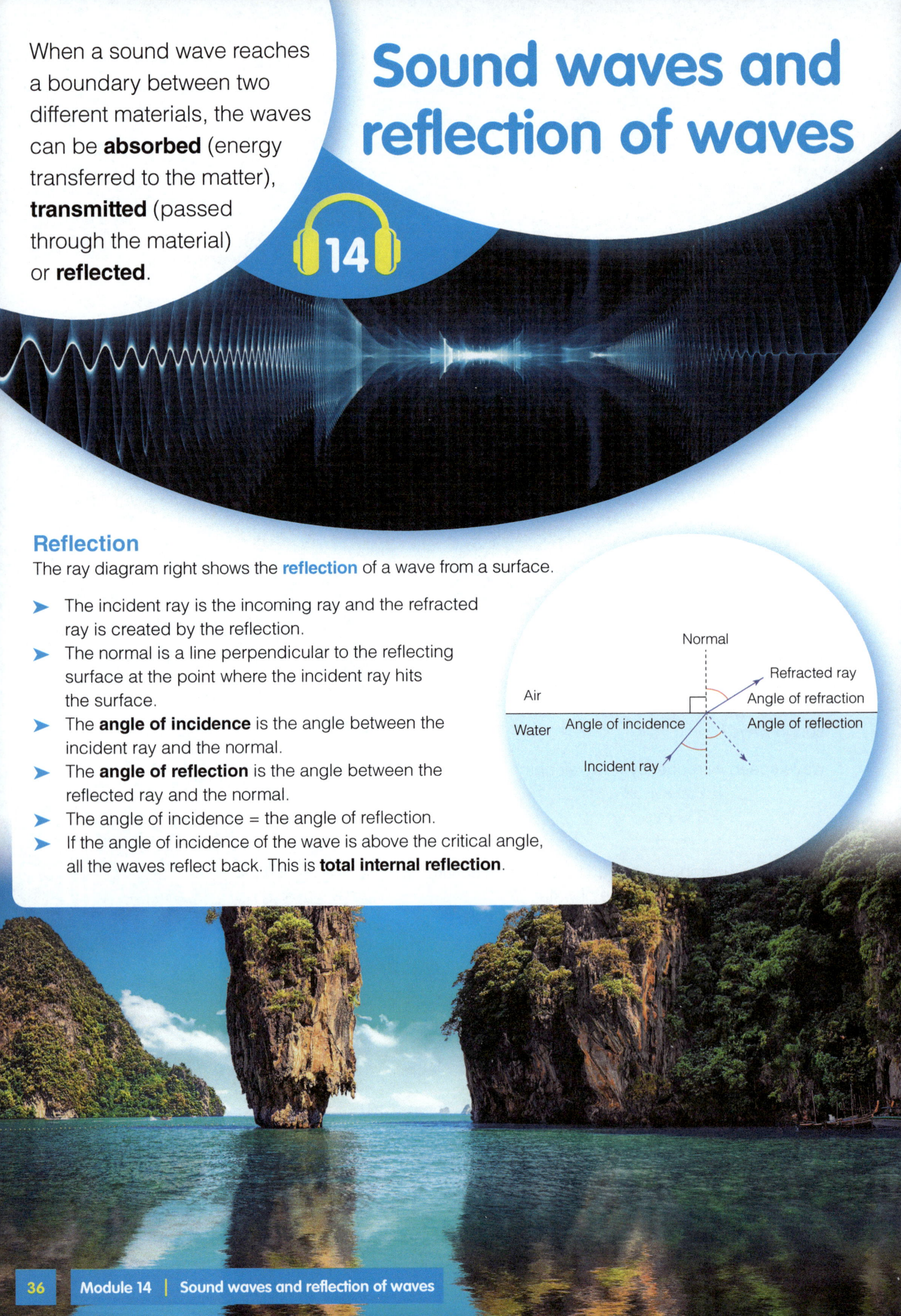

Normal
Air
Water
Refracted ray
Angle of refraction
Angle of incidence
Angle of reflection
Incident ray

Sound waves

Sound waves can travel through solids causing vibrations in the solid.

Within the ear, sound waves cause the ear drum and other parts to vibrate. These vibrations are transmitted to the brain by the nervous system and are perceived as sounds.

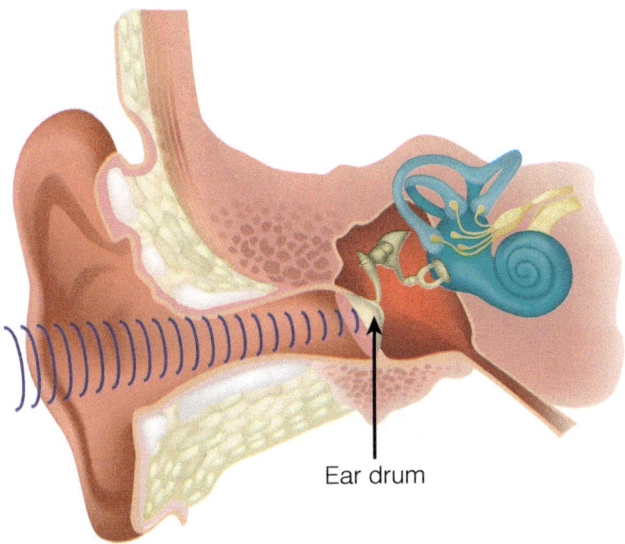

Ear drum

The conversion of sound waves to vibrations of solids works over a limited frequency range. Sounds outside this frequency range do not cause vibrations of the ear parts. As the ear parts do not vibrate, no signal is transmitted to the brain and therefore no sound is heard.

This means that the range of normal human hearing is from 20 Hz to 20 kHz. Sounds over 20 kHz are known as **ultrasound**. Sounds less than 20 Hz are **infrasound**. As a person ages, the human hearing range shrinks with a reduction in the upper range.

Keyword

Reflection ➤ Waves striking a boundary between different media and being returned back to the same media

Hold a small mirror vertical. Rotate the mirror until you can see yourself in it and then use a protractor to measure the angle the mirror has moved. Now repeat the experiment with the mirror a different distance from you. Does distance have an effect on the angle?

1. What does the angle of incidence equal?
2. Name a part of the ear that vibrates and allows sound to be perceived.
3. What is the normal frequency range of human hearing?

Waves for detection and exploration

Ultrasound

Ultrasound waves have a frequency higher than the upper limit of hearing for humans.

Ultrasound waves are partially reflected when they meet a boundary between two different media. The time taken for the reflections to reach a detector can be used to determine how far away such a boundary is. This allows ultrasound waves to be used for both medical and industrial imaging.

One of the major uses of ultrasound is prenatal scans.

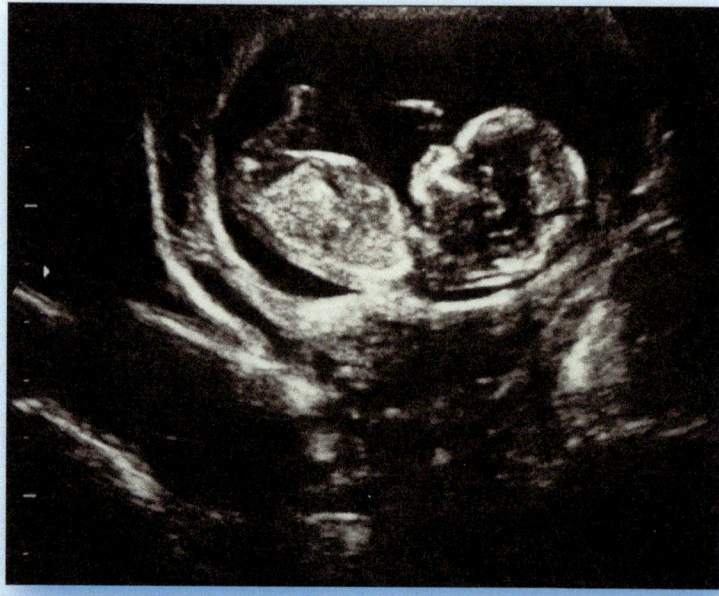

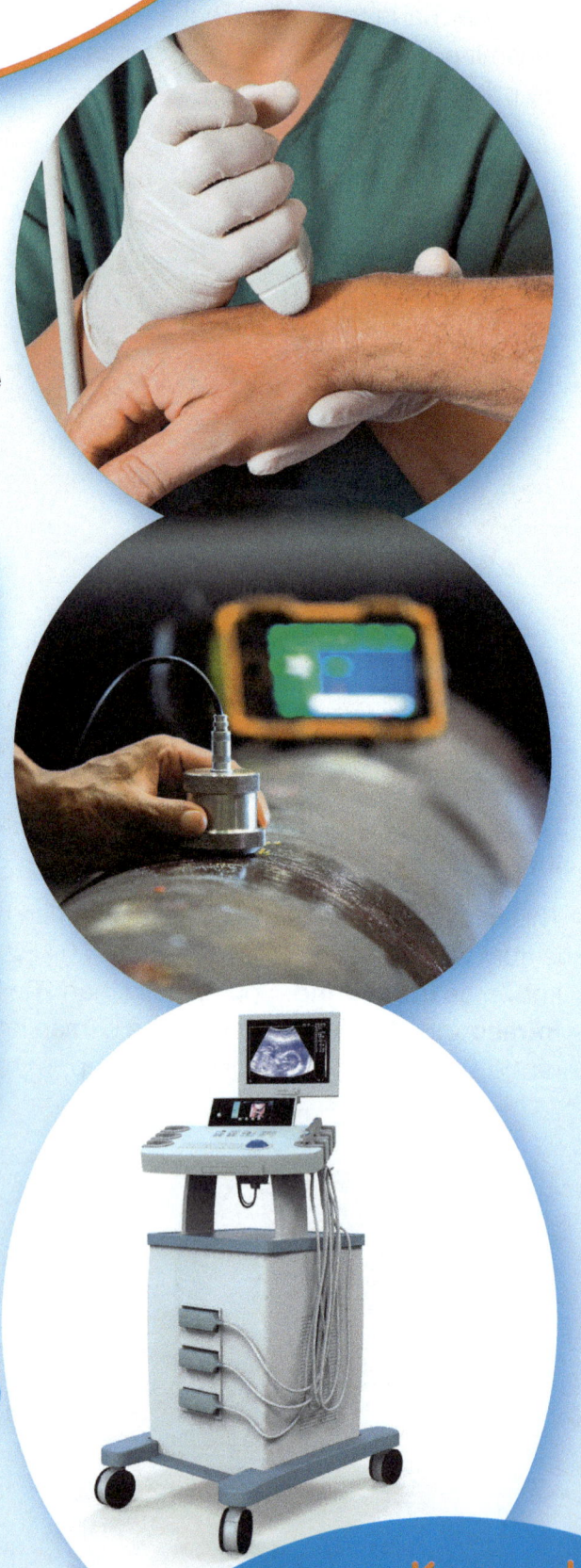

Write each of the wave types featured in this module on cards. On different cards write what they're used to detect. On a third set of cards write why the waves are suitable for this type of detection. Mix all the cards up and then rearrange them into the correct order.

Keyword

Ultrasound ➤ Longitudinal waves with a frequency higher than the upper limit of hearing for humans

Earthquakes

Earthquakes produce two types of seismic wave that travel through the Earth:

1. P-waves are longitudinal, seismic waves. P-waves travel at different speeds through solids and liquids.
2. S-waves are transverse, seismic waves. S-waves cannot travel through a liquid.

The P-waves and S-waves produced by an earthquake can be detected by a seismograph

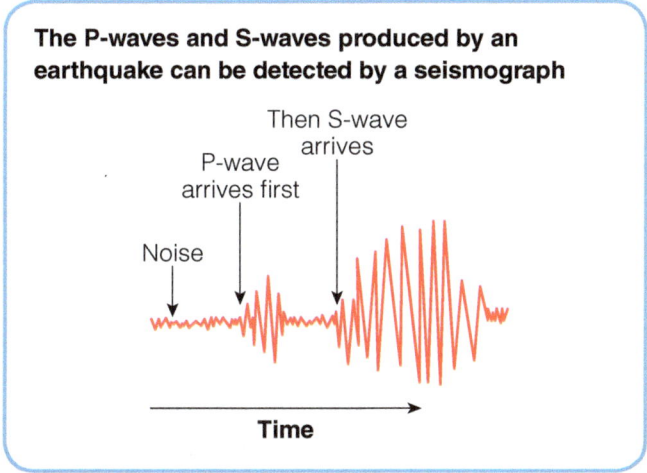

P-waves and S-waves have been used to provide evidence for the structure and size of the Earth's core.

As it is impossible to see the internal structure of the Earth, we must use models to study it.

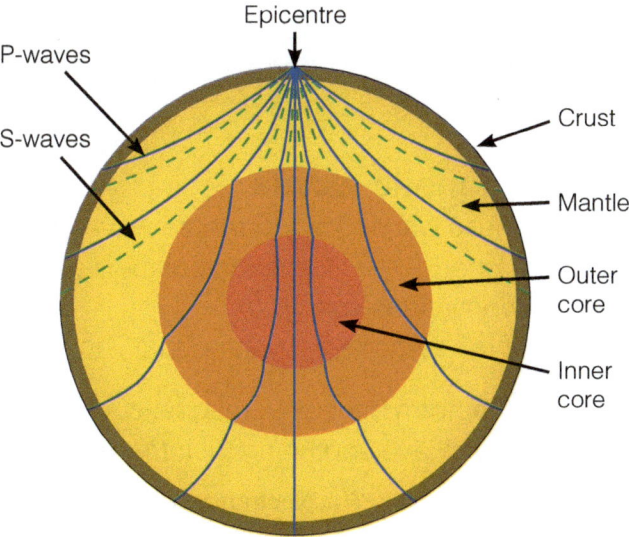

P-waves and S-waves are detected on the Earth's surface. Only P-waves are detected on the opposite side of the Earth from the epicentre of the earthquake, not S-waves. This helps prove that the outer core of the Earth is liquid, preventing the passage of S-waves.

Echo sounding

Echo sounding (sonar) using high frequency sound waves is used to detect objects in deep water and to measure water depth.

Fishing vessels use echo sounding to detect shoals of fish. The high frequency sound bounces off the group of fish and is detected by the boat. The time interval between the ultrasound being produced and detected allows the distance to the fish shoal to be determined.

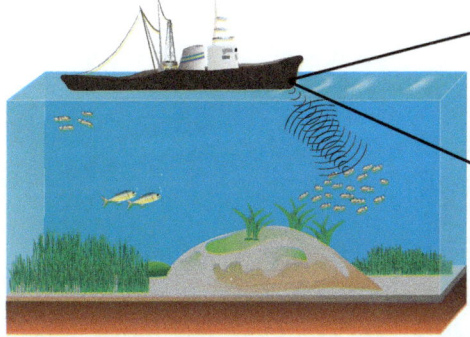

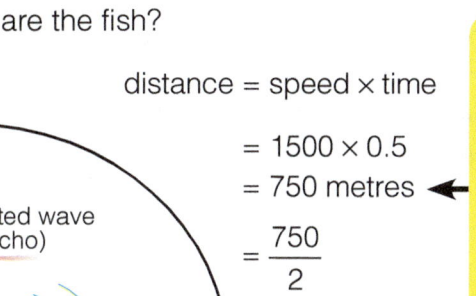

Example:

A fishing boat releases high frequency sound waves and detects the reflection from a shoal of fish 0.5 seconds later. The speed of sound in water is 1500 m/s. How far below the boat are the fish?

$$\text{distance} = \text{speed} \times \text{time}$$
$$= 1500 \times 0.5$$
$$= 750 \text{ metres}$$
$$= \frac{750}{2}$$
$$= 375 \text{ metres}$$

This value then needs to be divided by 2 as this is how far the sound waves must travel to the fish shoal and then back to the boat.

1. Give two differences between P-waves and S-waves.
2. Give one similarity and one difference between the use of ultrasound and echo sounding.
3. Why are S-waves not detected on the opposite point of the Earth's surface from an earthquake's epicentre?

Electromagnetic waves and properties 1

Electromagnetic waves

Electromagnetic waves are transverse waves that:

➤ transfer energy from the source of the waves to an absorber
➤ form a continuous spectrum
➤ travel at the same velocity through a vacuum (space) or air.

The waves that form the electromagnetic spectrum are grouped in terms of their wavelength and their frequency.

The electromagnetic spectrum

Radio waves	Micro-waves	Infrared radiation	Visible light	Ultraviolet	X-rays	Gamma rays

10^3m   1 m 10^{-3}m 10^{-5}m 10^{-7}m 10^{-9}m 10^{-11}m 10^{-13}m

Wavelength

Visible light

Human eyes only detect visible light so only identify a limited range of electromagnetic radiation.

Radio waves				Infrared	Ultra-violet	X-rays	Gamma rays
AM	FM	TV	Radar				
100 m	1 m	1 cm	0.01 cm	1000 nm	10 nm	0.01 nm	0.0001 nm

Visible light

Visible spectrum

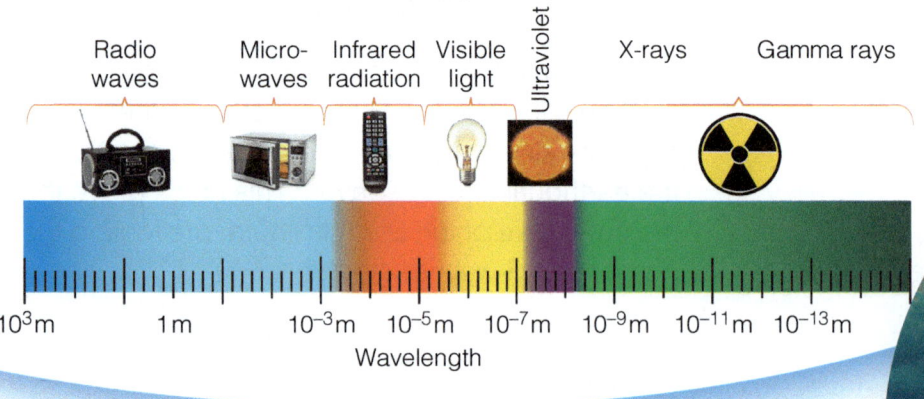

700 nm   600 nm   500 nm   400 nm

Wavelength

Different wavelengths of electromagnetic waves are reflected, refracted, absorbed or transmitted differently by different substances and types of surface.

Refraction

Refraction is when a wave changes direction as it travels from one medium to another.

Refraction occurs due to the wave changing speed as it moves between media. Waves have different velocities in different media.

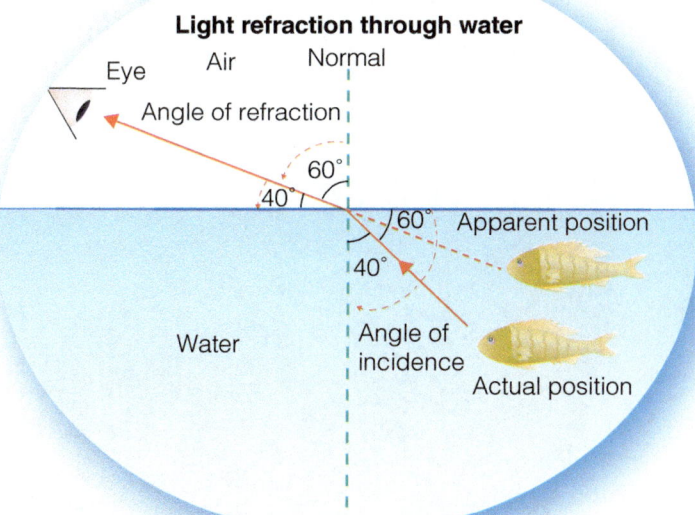

Light refraction through water

Keyword

HT **Refraction** ➤ Change in direction of a wave as it travels from one medium to another

Light rays travelling through water

In the example above, the fish appears to be above its actual position due to the refraction of the light rays. Light travels faster in the air than it does in the water. This leads to the light ray bending away from the normal. When light slows down, it bends towards the normal.

The wave front diagram below shows a wave moving from air (less optically dense) to water (more optically dense). As the wave travels slower in a denser medium, the edge of the wave that hits the water first slows down whilst the rest of the wave continues at the same speed. This causes the light to bend towards the normal. The opposite effect occurs when a wave moves from a more optically dense medium to a less optically dense medium.

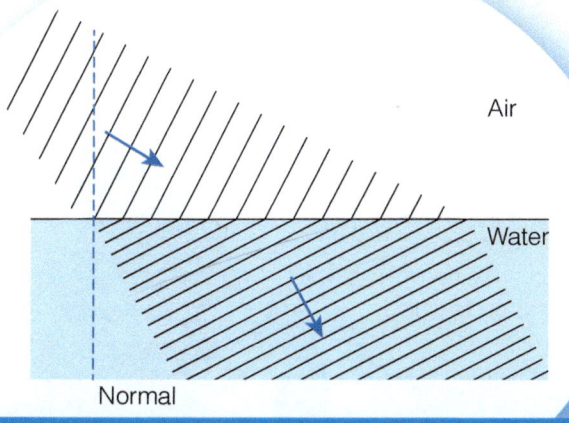

Write each of the different types of electromagnetic radiation on separate cards. Mix them up and then rearrange them into the correct order, from largest wavelength to smallest wavelength.

1. Which has the largest wavelength: infrared radiation or ultraviolet radiation?
2. What section of electromagnetic radiation can be perceived by the human eye?
HT 3. Which way do waves bend when they enter a more optically dense medium?

Electromagnetic waves and properties 2

Keyword
Sievert ➤ Unit of radiation dose

17

Electromagnetic waves

Changes in atoms and the nuclei of atoms can result in electromagnetic waves being generated or absorbed over a wide frequency range.

Gamma rays originate from changes in the nucleus of an atom.

HT Radio waves

Radio waves can be produced by oscillations in electrical circuits. When radio waves are absorbed they may create an alternating current with the same frequency as the radio wave. Therefore, radio waves induce oscillations in an electrical circuit.

The diagram below shows the main features of electromagnetic waves.

The effects of electromagnetic waves depend on the type of radiation and the size of the dose.

Ultraviolet waves, x-rays and gamma rays can have hazardous effects on human body tissue.

Radiation dose (in **sieverts**) is a measure of the risk of harm from an exposure of the body to the radiation.

1000 millisieverts (mSv) equals 1 sievert (Sv)

Microwaves can cause internal heating of body cells.

Ultraviolet waves can cause skin to age prematurely and increase the risk of skin cancer.

X-rays and gamma rays are ionising radiation that can cause the mutation of genes and cancer.

Make separate cards of the names of each electromagnetic radiation, its uses and any associated dangers. Mix up all these cards and then match them up.

Uses of electromagnetic waves

Electromagnetic waves have many practical applications.

Radio waves	Television, radio, bluetooth	Low frequency radio waves can diffract around hills and reflect off the ionosphere, meaning they don't require a direct line of sight between transmitter and receiver.
Microwaves	Satellite communications, cooking food	Their small wavelength allows them to be directed in narrow beams.
Infrared	Electrical heaters, cooking food, infrared cameras	Thermal radiation heats up objects.
Visible light	Fibre optic communications	This can be reflected down a fibre optic cable.
Ultraviolet	Energy efficient lamps, sun tanning	Ultraviolet is used for low energy light bulbs to produce white light. This requires less energy than filament bulbs.
X-rays	Medical imaging and treatments	X-rays are absorbed differently by different parts of the body; more are absorbed by hard tissues, such as bone, and less are absorbed by soft tissues. This allows images of the inside of the body to be created.
Gamma rays	Sterilising, medical imaging and treatment of cancer	Gamma rays destroy living cells so can be used to sterilise medical equipment and apparatus. Gamma rays can also be used to destroy cancerous tumours and carry out functional organ scans.

HT

HT 1. How are radio waves produced?
2. What are the potential dangers of gamma rays?
3. Give one use of ultraviolet radiation.

Lenses and visible light

Lenses

A lens forms an image by refracting light.

➤ Convex lens – parallel rays of light are brought to a focus at the principal focus. Convex lenses can form virtual or real images.

➤ Concave lens – parallel rays of light diverge when they pass through a concave lens. Concave lenses always form virtual images.

Real images and virtual images

The distance from the lens to the principal focus is called the **focal length**.

Distance of object from convex lens	Image
More than 2 × focal length	Real and diminished
2 × focal length	Real and same size as object
Between 2 × focal length and the focal length	Real and magnified
Closer than focal length	Virtual

The effect of convex and concave lenses can be shown by ray diagrams. The lenses will be represented by symbols.

More powerful lenses have a shorter focal length. Thicker convex lenses are generally more powerful than thinner ones. The magnification produced by a lens can be calculated using the following equation:

$$\text{magnification} = \frac{\text{image height}}{\text{object height}}$$

Image heights are usually measured in mm or cm.

Example:

What is the magnification of an image that is 8 cm high when the original object was 0.3 cm high?

$$\text{magnification} = \frac{\text{image height}}{\text{object height}}$$

$$= \frac{8}{0.3}$$

$$= 27 \times \text{magnification}$$

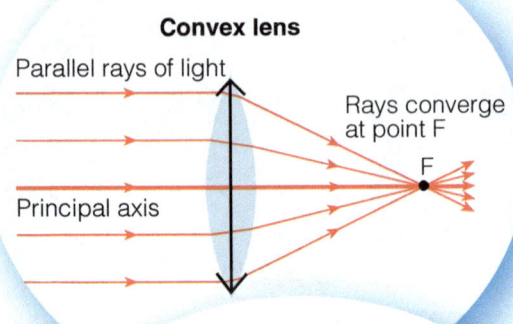

Convex lens

Parallel rays of light

Rays converge at point F

F

Principal axis

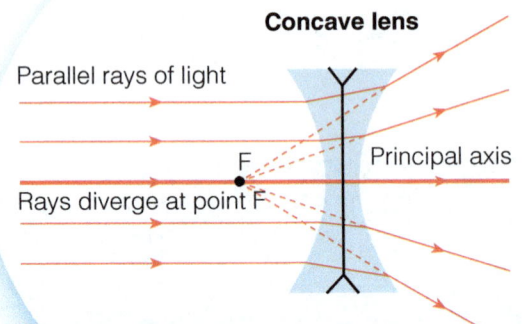

Concave lens

Parallel rays of light

F

Principal axis

Rays diverge at point F

Keywords

Opaque ➤ Object that reflects light

Transparent ➤ Object that allows all light to pass through

Translucent ➤ Object that diffuses the light that passes through it, meaning that things observed through a translucent material are not properly visible

Visible light

Each colour within the visible light spectrum has a narrow range of wavelengths and frequency.

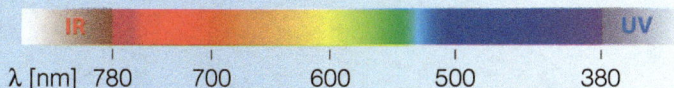

λ [nm]   780   700   600   500   380

Opaque objects reflect or absorb all light incident on them. The colour of an opaque object is determined by which wavelengths of light are more strongly reflected. Wavelengths not reflected are absorbed.

For example, a blue object absorbs all wavelengths of visible light except blue. The blue wavelength is diffusely reflected into the eyes of the observer.

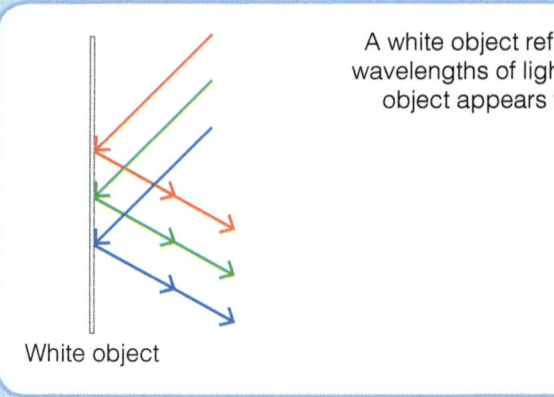

A white object reflects all wavelengths of light, so the object appears white.

White object

A black object absorbs all wavelengths of light (no reflection occurs).

Black object

A blue object absorbs all wavelengths of light except blue which it reflects. It therefore appears blue.

Blue object

There are two types of reflection:

➤ Specular reflection – reflection from a smooth surface in a single direction, e.g. reflection from a mirror.
➤ Diffuse reflection – reflection from a rough surface causes scattering. Diffuse reflection allows us to see objects.

Objects that transmit light are either **transparent** (allow all light to pass through) or **translucent** (allow light through but diffuses it, causing things observed through a translucent object to not be properly visible).

Colour filters are transparent objects that transmit certain wavelengths of light and absorb all other wavelengths of light. For example, a red filter transmits red light and absorbs all other visible light wavelengths.

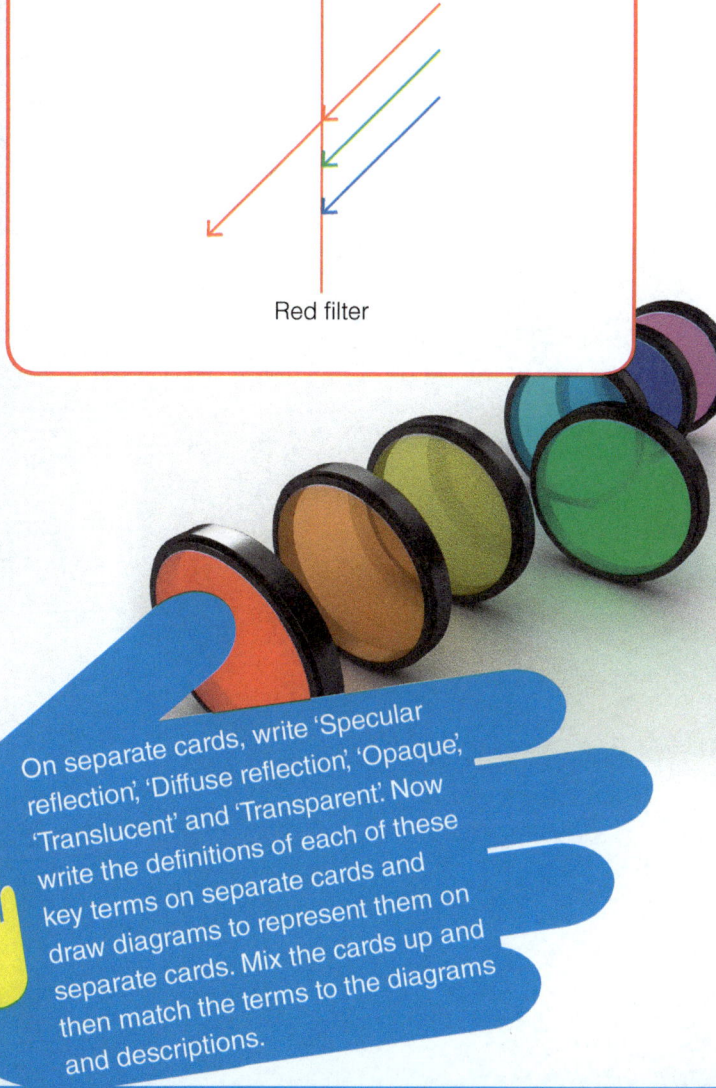

Red filter

On separate cards, write 'Specular reflection', 'Diffuse reflection', 'Opaque', 'Translucent' and 'Transparent'. Now write the definitions of each of these key terms on separate cards and draw diagrams to represent them on separate cards. Mix the cards up and then match the terms to the diagrams and descriptions.

1. What type of image is formed by a concave lens?
2. What light is transmitted by a blue filter?
3. What is the original height of an object that has been magnified 4 times to produce a 17 cm high image?

Black body radiation

Keyword

Emission ➤ Release of energy

Emission and absorption of infrared radiation

All bodies **emit** and absorb infrared radiation. The hotter the body, the more infrared radiation it emits in a given time.

A perfect black body is a theoretical object that:

➤ absorbs all of the radiation incident on it
➤ does not reflect or transmit any radiation
➤ is the best possible emitter as it emits the maximum amount of radiation possible at a given temperature.

A perfect black body is impossible in the real world as no object can fulfil these criteria.

All bodies emit radiation, and the intensity and wavelength distribution of any emission depends on the temperature of the body.

Objects with a high temperature appear white hot, like this molten metal.

A body at a constant temperature is absorbing radiation at the same rate as it is emitting radiation. The temperature of a body increases when the body absorbs radiation faster than it emits radiation.

The temperature of the Earth depends on many factors including:

➤ the rates of absorption of radiation
➤ the rate of emission of radiation
➤ the rate of reflection of radiation into space.

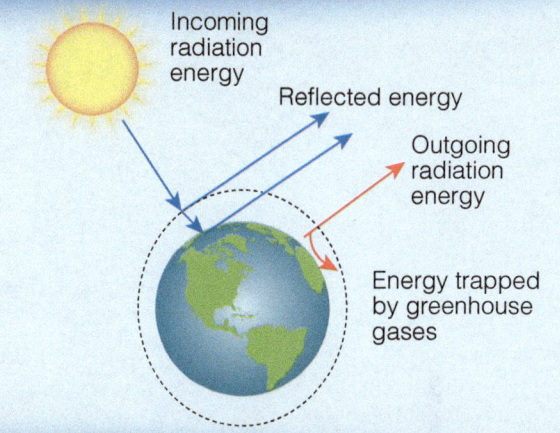

Incoming radiation energy

Reflected energy

Outgoing radiation energy

Energy trapped by greenhouse gases

The temperature of a body is related to the radiation absorbed and the radiation emitted.

For example, when food is beneath a grill it is absorbing more radiation than it is emitting so its temperature is increased. When the food is removed from the grill, it now emits more radiation than it is absorbing. This means the food cools down.

Create a large flow diagram showing energy being absorbed, released and emitted from the Earth. Include as many energy sources as you can.

1. Give the three main characteristics of a perfect black body.
2. Give two factors the temperature of the Earth depends on.
3. What will happen to the temperature of a body that is emitting as much radiation as it is absorbing?

Mind map

Echo sounding

Ultrasound

Compression

Sound waves

Rarefaction

Wave speed

P-waves

Longitudinal

Refraction

WAVES

Reflection

Hazards

Transverse

Electromagnetic spectrum

Uses

S-waves

Infrared radiation

Visible light

Perfect black body

Lenses

Convex

Concave

Practice questions

1. How does compression differ from rarefaction? **(2 marks)**

2. What is the frequency of a wave travelling at 200 m/s with a wavelength of 3.2 m? **(2 marks)**

3. Explain how P-waves and S-waves from an earthquake can be used to draw conclusions about the structure of the Earth. **(4 marks)**

4. A 5 cm high image is formed by a magnification of 2.2.

   What is the height of the original object? **(2 marks)**

5. Look at the ray diagram below.

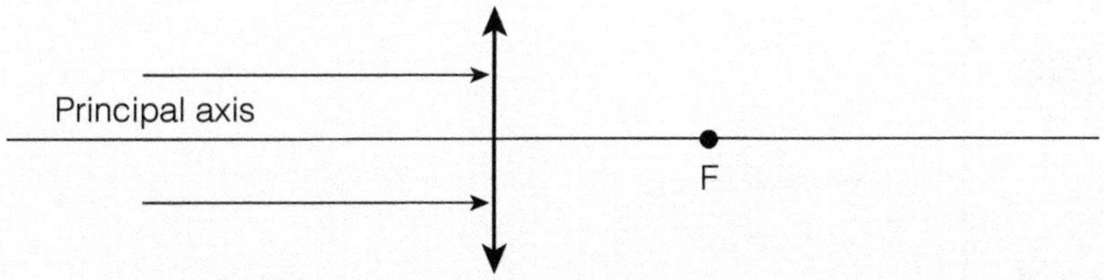

   Principal axis

   F

   a) What type of lens is shown in the diagram above? **(1 mark)**

   b) Complete the ray diagram to show the effect of the lens. **(1 mark)**

6. What does the colour of green plants suggest about what light is reflected and absorbed by them? **(2 marks)**

7. David wrote the following explanation of how sound travels:

   'Sound travels as a transverse wave. When a sound wave travels in air the oscillations of the air particles are perpendicular to the direction of energy transfer. The sound wave transfers matter and energy.'

   Write this paragraph out again, correcting the three mistakes David has made. **(3 marks)**

8. Explain the difference between the terms infrasound and ultrasound. **(2 marks)**

9. How can sonar be used to detect shoals of fish? **(3 marks)**

Circuits, charge and current

Circuit symbols

The diagram below shows the standard symbols used for components in a circuit.

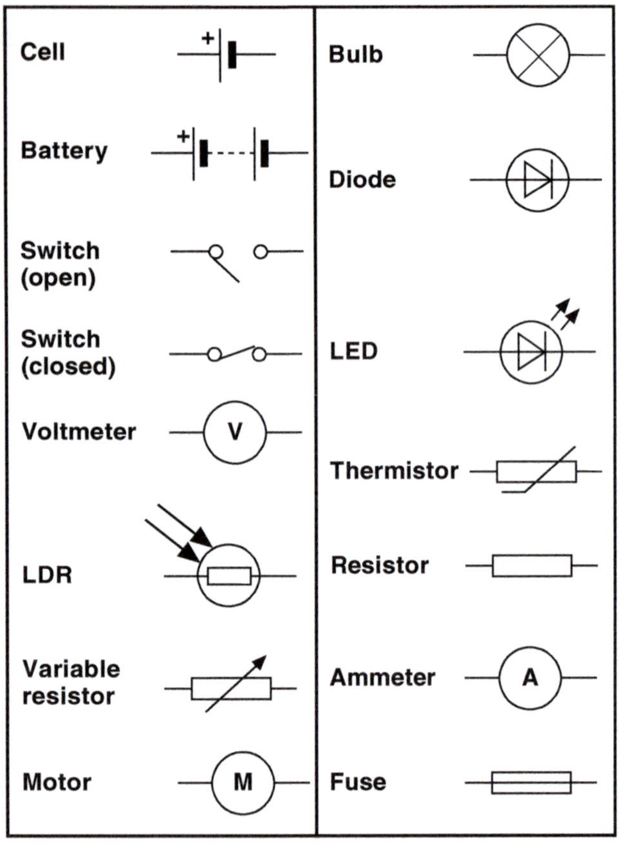

Here is an example of a circuit diagram:

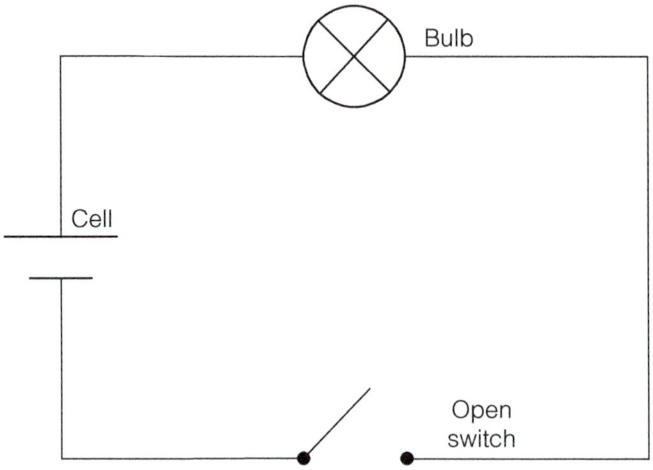

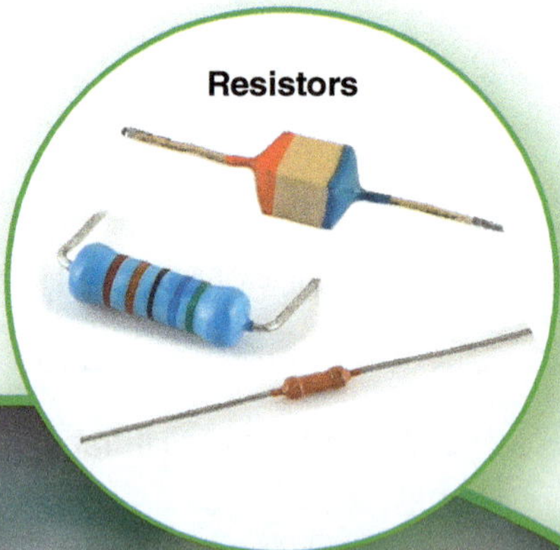

WS

Cell	—+⊣⊢—	Bulb	—⊗—
Battery	—+⊣⊢- - -⊣⊢—	Diode	—◁|—
Switch (open)	—o ⁄o—	LED	—◁|↗↗—
Switch (closed)	—o⁄ o—	Thermistor	—▭—
Voltmeter	—(V)—	Resistor	—▭—
LDR	↘↘(▭)		
Variable resistor	—▭↗—	Ammeter	—(A)—
Motor	—(M)—	Fuse	—▭—

Bulb

Cell

Open switch

Resistors

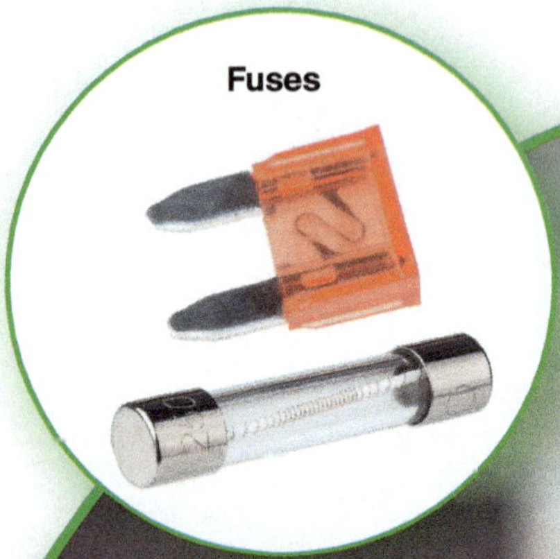

Fuses

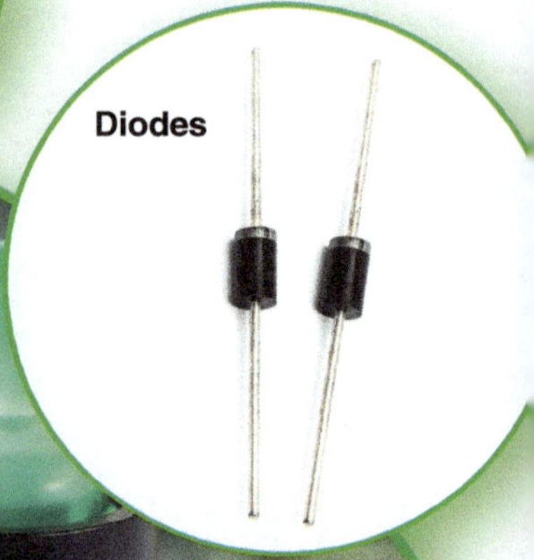

Diodes

Switches

AMPERES

Electrical charge and current

For electrical charge to flow through a closed circuit, the circuit must include a source of energy that produces a potential difference, such as a battery, cell or powerpack.

Electric current is a flow of electrical charge. The size of the electric current is the rate of flow of electrical charge. Charge flow, current and time are linked by the following equation:

charge flow = current × time

$$Q = It$$

➤ charge flow, Q, in coulombs, C
➤ current, I, in amperes or amps A
➤ time, t, in seconds, s

Example:
A current of 6 A flows through a circuit for 14 seconds. What is the charge flow?

charge flow = current × time
= 6 × 14
= 84 C

The current at any point in a single closed loop of a circuit has the same value as the current at any other point in the same closed loop.

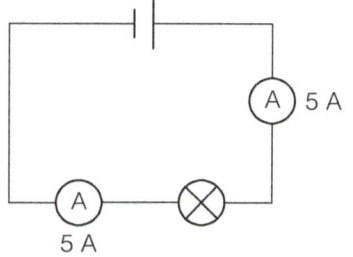

Both ammeters in the circuit above show the same current of 5 amps.

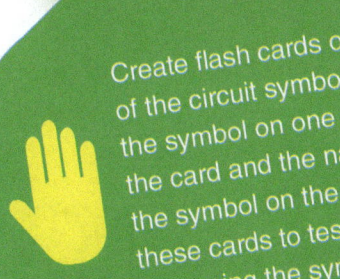

Create flash cards of each of the circuit symbols. Draw the symbol on one side of the card and the name of the symbol on the other. Use these cards to test yourself at naming the symbols.

1. What is the difference between the symbol for a resistor and the symbol for a variable resistor?
2. What charge flows through a circuit per second if the current is 4.2 A?
3. What is the current of a charge flow of 450 C in five seconds?

20

Current, resistance and potential difference

Current, resistance and potential difference

The current through a component depends on both the resistance of the component and the potential difference (p.d.) across the component. Potential difference is the energy transferred per unit charge passed.

The greater the resistance of the component, the smaller the current for a given potential difference across the component.

Current, potential difference or resistance can be calculated using the following equation:

> **potential difference = current × resistance**
> $$V = IR$$
> ➤ potential difference, *V*, in volts, V
> ➤ current, *I*, in amperes or amps, A
> ➤ resistance, *R*, in ohms, Ω

Example:
A 5 ohm resistor has a current of 2 A flowing through it. What is the potential difference across the resistor?

potential difference = current × resistance
$$= 5 \times 2$$
$$= 10 \text{ V}$$

By measuring the current through, and potential difference across a component, it's possible to calculate the resistance of a component.

The circuit diagram below would allow you to determine the resistance of the filament lamp.

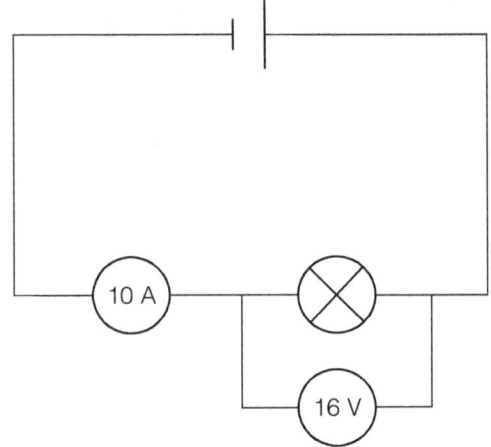

potential difference = current × resistance

$$\text{resistance} = \frac{\text{potential difference}}{\text{current}}$$

$$= \frac{16}{10}$$

$$= 1.6 \text{ ohms}$$

Create a circuit using coloured wool or string – include at least six different components. Stick it onto a large sheet of paper and then annotate it.

Keyword

Ohmic conductor ➤ Resistor at constant temperature where the current is directly proportional to the potential difference

Resistors

In an **ohmic conductor**, at a constant temperature the current is directly proportional to the potential difference across the resistor. This means that the resistance remains constant as the current changes.

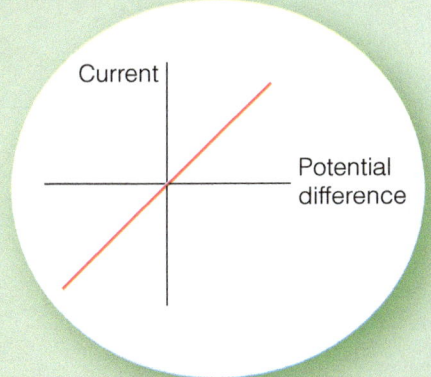

The resistance of components such as lamps, diodes, thermistors and LDRs is not constant; it changes with the current through the component. They are not ohmic conductors.

When a current flows through a resistor, the energy transfer causes the resistor to heat up. This is due to collisions between electrons and the ions in the lattice of the resistor.

This heating can be an advantage, such as in an electrical heater. It is also a disadvantage as it can lead to electrical devices being damaged due to overheating. Thicker wires have a lower resistance as there is a larger cross-sectional area for the current to pass through.

Filament lamps

The resistance of a filament lamp increases as the temperature of the filament increases.

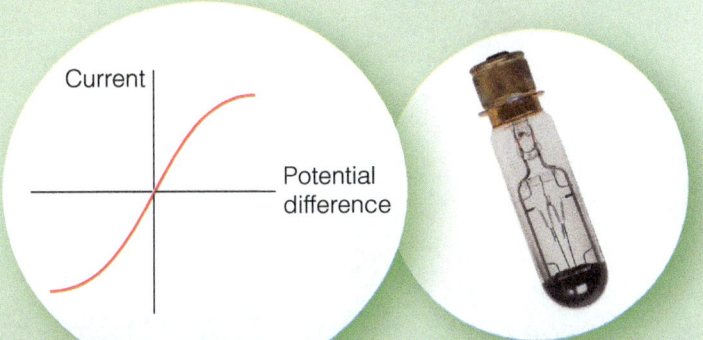

Diodes

The current through a diode flows in one direction only. This means the diode has a very high resistance in the reverse direction.

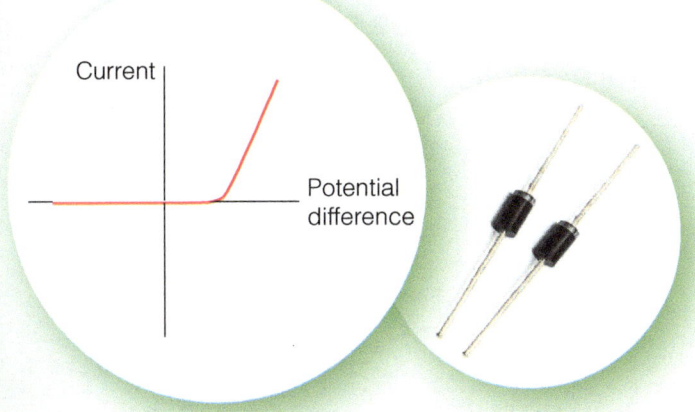

Light dependent resistors (LDR)

The resistance of an LDR decreases as light intensity increases. LDRs are used in circuits where lights are required to switch on when it gets dark, such as floodlights. When the resistance decreases due to the lack of light, sufficient current flows through the LDR to activate the light.

Thermistors

The resistance of a thermistor decreases as the temperature increases. Thermistors are used in thermostats to control heating systems.

21

1. Explain why an LDR is not an ohmic conductor.
2. What two pieces of equipment need to be wired into a circuit in order to determine the resistance of a component in the circuit?
3. What is the potential difference if the current is 4 A and the resistance is 2 ohms?
4. Calculate the resistance of a component that has a current of 3 A flowing through it and a potential difference of 6 V.

Components can be joined together in either a **series circuit** or a **parallel circuit**. Some circuits can include both series and parallel sections.

Series and parallel circuits

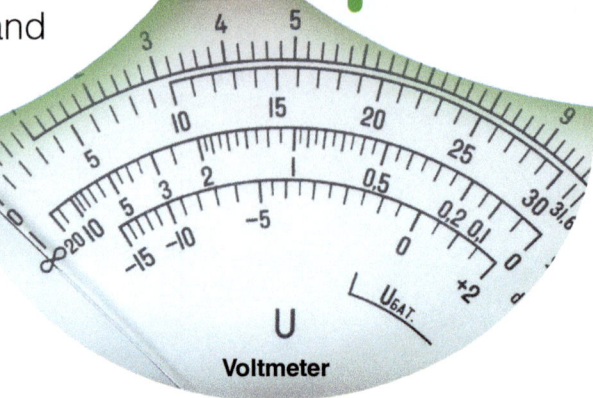

Voltmeter

Series circuits

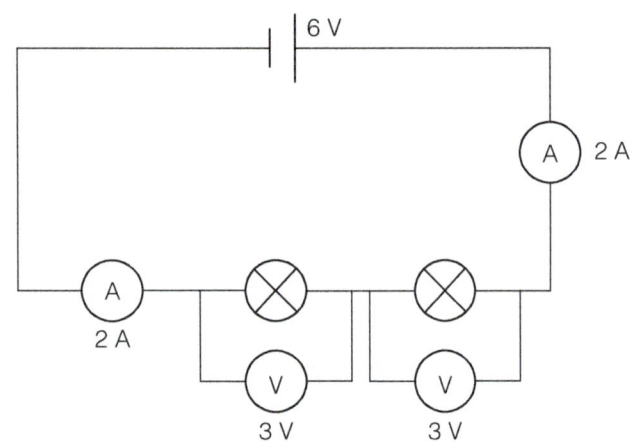

For components connected in series:

➤ there is the same current through each component
➤ the total potential difference of the power supply is shared between the components
➤ the total resistance of two components is the sum of the resistance of each component.

Total resistance is given by the following equation:

> **R total = R1 + R2**
> ➤ resistance, *R*, in ohms, Ω

Example:
What is the total resistance of the two resistors in the series circuit below?

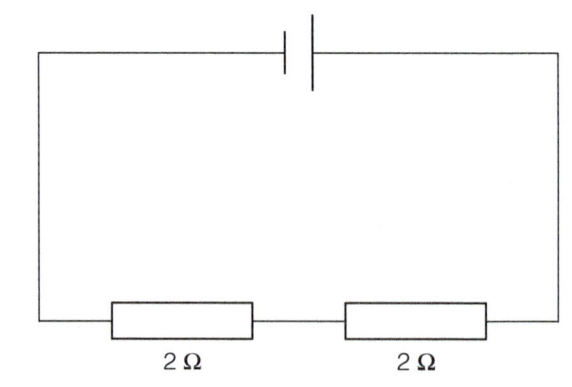

Total = R1 + R2
= 2 Ω + 2 Ω
= 4 Ω

Keywords

Series circuit ➤ Circuit where all components are connected along a single path
Parallel circuit ➤ Circuit which contains branches and where all the components will have the same voltage

Parallel circuits

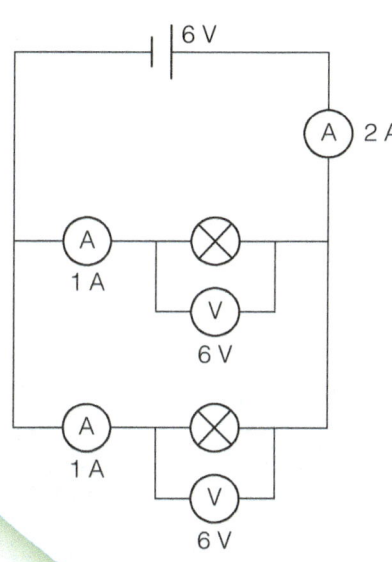

For components connected in parallel:

➤ the potential difference across each component is the same

➤ the total current through the whole circuit is the sum of the currents through the separate components. The current splits between the branches of the circuit and combines when the branches meet

➤ the total resistance of two resistors is less than the resistance of the smallest individual resistor. This is due to the potential difference across the resistors being the same but the current splitting.

22

Draw a parallel circuit of your own design, making sure that you include at least two resistors.

1. If the current through one component in a series circuit is 5 A, what is the current through the rest of the components?
2. How do you calculate the total resistance of two resistors in series?
3. A 3 Ω resistor and a 2 Ω resistor are connected in parallel. Would the resistance be higher or lower than 2 Ω?

Domestic uses and safety

Direct and alternating current

Cells and batteries supply current that always passes in the same direction. This is direct current (**dc**).

Alternating current (**ac**) changes direction at a frequency of fifty times a second. Mains electricity is an ac supply. In the UK it has a frequency of 50 Hz and is about 230 V.

Mains electricity

WS Most electrical appliances are connected to the mains using a three-core cable with a three-pin plug.

Live wire	Brown	Carries the alternating potential difference from the supply.
Neutral wire	Blue	Completes the circuit. The neutral wire is at, or close to, earth potential (0 V).
Earth wire	Green and yellow stripes	The earth wire is at 0 V. It only carries a current if there is a fault.

The potential difference between the live wire and earth (0 V) is about 230 V.

Our bodies are at earth potential (0 V). Touching a live wire produces a large potential difference across our body. This causes a current to flow through our body, resulting in an electric shock that could cause serious injury or death.

Insulation, fuses and circuit breakers

If an electrical fault causes too great a current, the circuit is disconnected by a fuse or a circuit breaker connected to the live wire.

The current will cause the fuse to overheat and melt or the circuit breaker to switch off (trip). A circuit breaker operates much faster than a fuse and can be reset.

Appliances with metal cases are usually earthed. If a fault occurs, a large current flows from the live wire to earth. This melts the fuse and disconnects the live wire.

Some appliances are double insulated meaning it is impossible for the case to become live. (Either the case is plastic or it is impossible for the live wire to come into contact with the casing). Double insulated appliances have no earth connection.

Electric drills are examples of appliances that are double insulated.

Keywords

dc ➤ Direct current that always passes in the same direction

ac ➤ Alternating current that changes direction

Circuit breaker

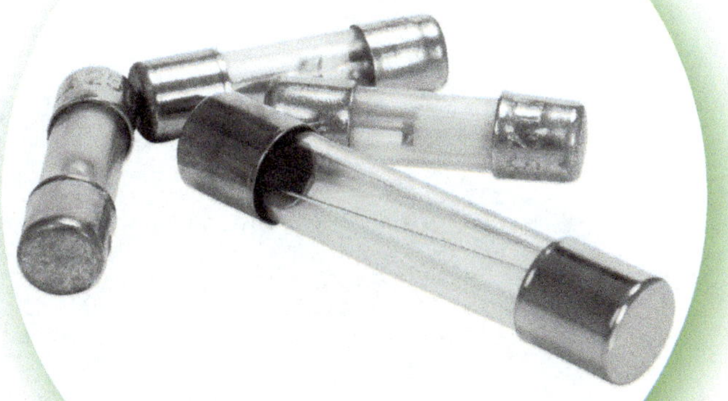

Fuses

Make small cards of the colours, names and functions of the wires in a three-pin plug. Mix the cards up and then arrange them to show the correct colours, functions and names of the wires.

1. Why don't double insulated appliances have an earth connection?
2. What are the advantages of a circuit breaker over a conventional fuse?
3. Why is it dangerous to touch a live wire?

Keyword

National Grid ➤ System of transformers and cables linking power stations to consumers

Power

The power of a device is related to the potential difference across it and the current through it by the following equations:

> **power = potential difference × current**
>
> $$P = VI$$
>
> **or**
>
> **power = current² × resistance**
>
> $$P = I^2 R$$
>
> ➤ power, P, in watts, W
> ➤ potential difference, V, in volts, V
> ➤ current, I, in amperes or amps, A
> ➤ resistance, R, in ohms, Ω

Example:

A bulb has a potential difference of 240 V and a current flowing through it of 0.6 A. What is the power of the bulb?

power = potential difference × current
= 240 × 0.6
= 144 W

Energy transfers in everyday appliances

Everyday electrical appliances are designed to bring about energy transfers.

The amount of energy an appliance transfers depends on how long the appliance is switched on for and the power of the appliance.

Here are some examples of everyday energy transfer in appliances:

➤ A hairdryer transfers electrical energy from the ac mains to kinetic energy (in an electric motor to drive a fan) and heat energy (in a heating element).
➤ A torch transfers electrical energy from batteries into light energy from a bulb.

Work done

Work is done when charge flows in a circuit.

The amount of energy transferred by electrical work can be calculated using the following equation:

energy transferred = power × time

$$E = Pt$$

and

energy transferred = charge flow × potential difference

$$E = QV$$

- ➤ energy transferred, E, in joules, J
- ➤ power, P, in watts, W
- ➤ time, t, in seconds, s,
- ➤ charge flow, Q, in coulombs, C
- ➤ potential difference, V, in volts, V

The National Grid

The **National Grid** is a system of cables and transformers linking power stations to consumers.

Electrical power is transferred from power stations to consumers using the National Grid.

Step-up transformers **increase** the potential difference from the power station to the transmission cables.

Step-down transformers **decrease** the potential difference to a much lower and safer level for domestic use.

Increasing the potential difference reduces the current so reduces the energy loss due to heating in the transmission cables. Reducing the loss of energy through heat makes the transfer of energy much more efficient. Also, the wires would glow and be more likely to break over time if the current through them was high.

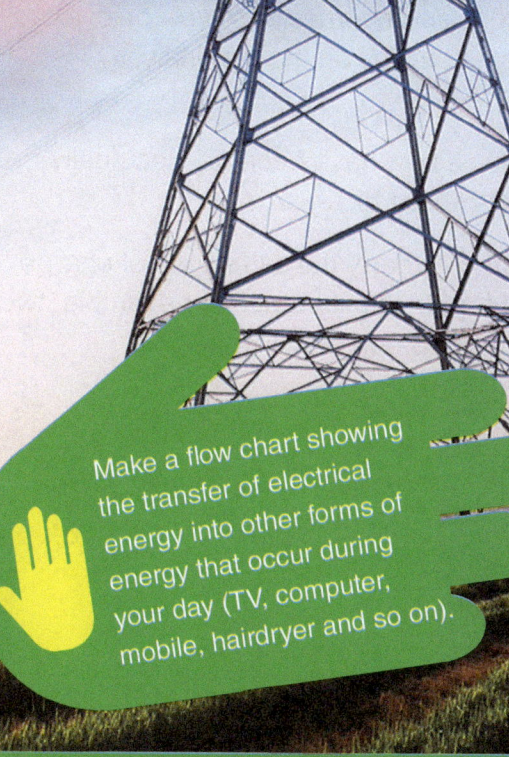

Make a flow chart showing the transfer of electrical energy into other forms of energy that occur during your day (TV, computer, mobile, hairdryer and so on).

1. What is the power of a device that has a current flowing through it of 5 A and a resistance of 2 Ω?
2. What is the energy transferred by a charge flow of 60 C and a potential difference of 12 V?
3. Why are step-down transformers important in the National Grid?

Static electricity

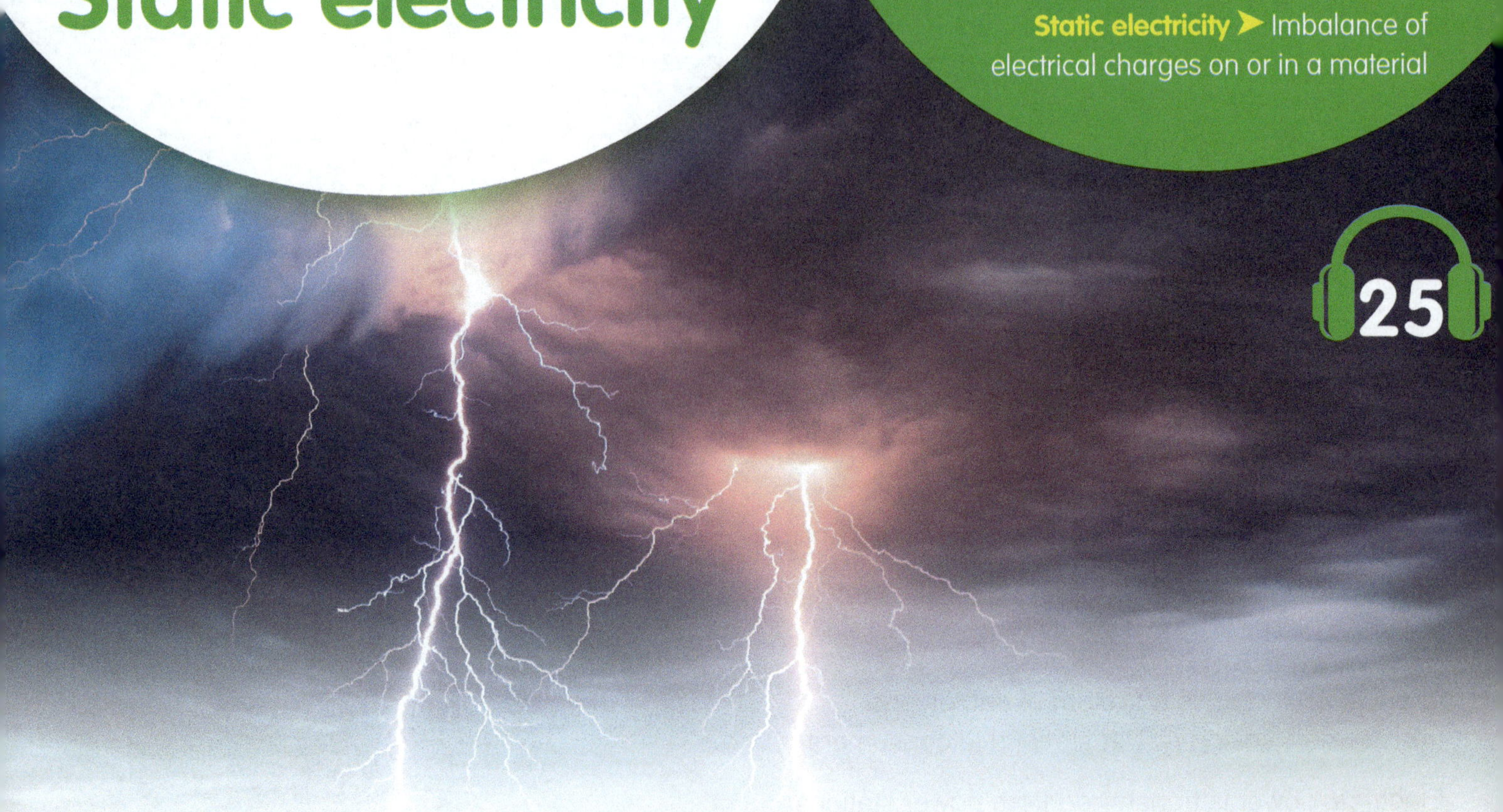

Static charge

When some insulating materials are rubbed against each other, they become electrically charged. One material gains electrons from the other and becomes negatively charged. The other material loses electrons so becomes positively charged. As the charge remains on the materials, this is **static electricity**.

These materials now have become electrically charged. This is a static charge.

If two objects that carry the same type of charge are brought together, they repel. For example, two positively charged objects will repel.

Two objects that carry different types of charge will attract. For example, a positively charged object will attract a negatively charged object.

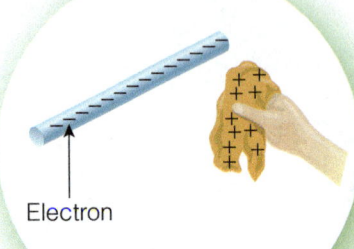

Electron

Attraction and repulsion between two charged objects are examples of non-contact forces.

The greater the charge on an isolated object, the greater the potential difference between the object and earth.

If this potential difference becomes high enough, a spark may jump across the gap between the object and an earthed conductor brought near to it.

Examples of static electricity include:
➤ electric shocks from everyday objects
➤ lightning
➤ a charged balloon attached to a wall
➤ a charged comb picking up small pieces of paper.

Earthing can be used to remove excess charge by allowing the electrons to flow down the earthed conductor. This is important in preventing dangers of sparking, such as when fuelling cars where a spark could lead to fuel igniting.

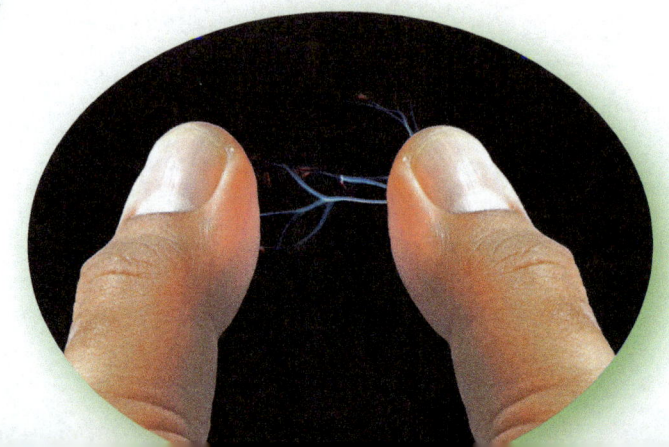

Electric fields

A charged object creates an electric field around itself. An electrical field is a region where an electrical charge experiences a force. The electric field is strongest close to the charged object. The further away from the charged object, the weaker the field. The number and density of field lines show the strength of the field – the more lines, the stronger the field.

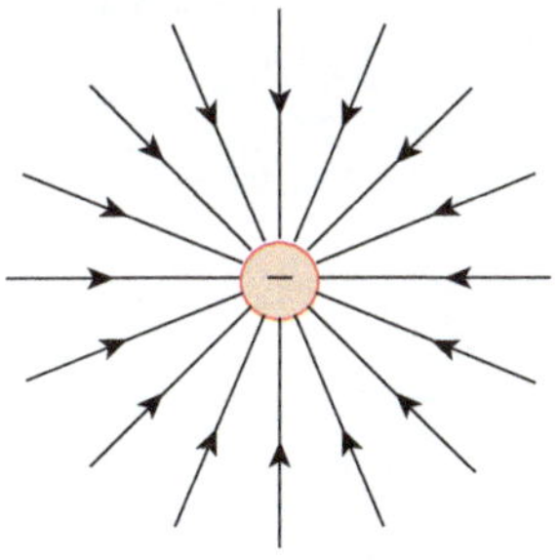

A second charged object placed in the field experiences a force. The force gets stronger as the distance between the objects decreases.

This force is the attraction and repulsion felt by charged objects when they are brought together. The field lines show the direction the force would act on a positively charged object. The electrical field is also what leads to the sparking between charged objects.

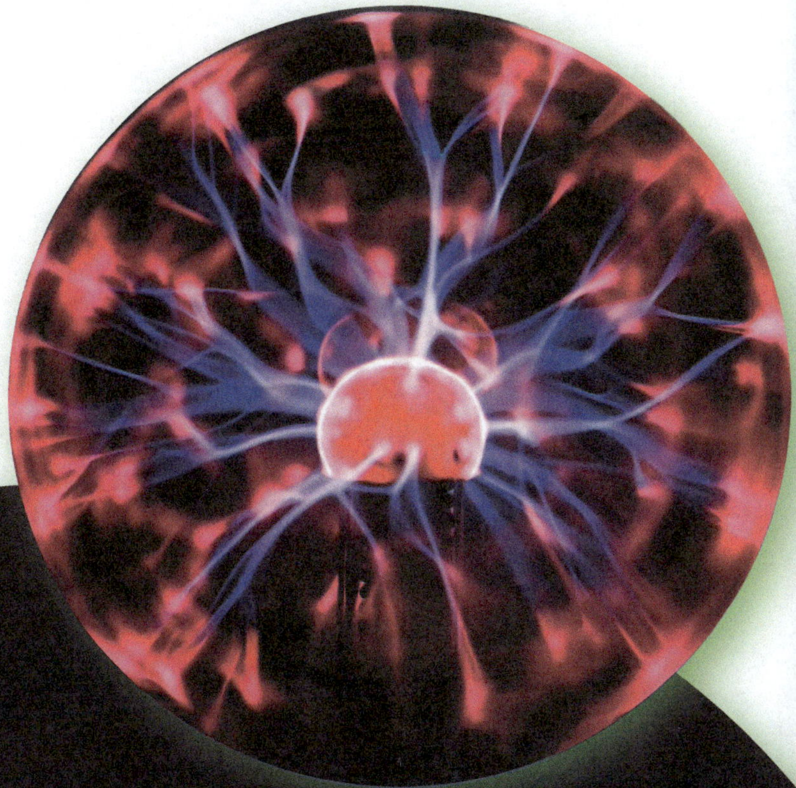

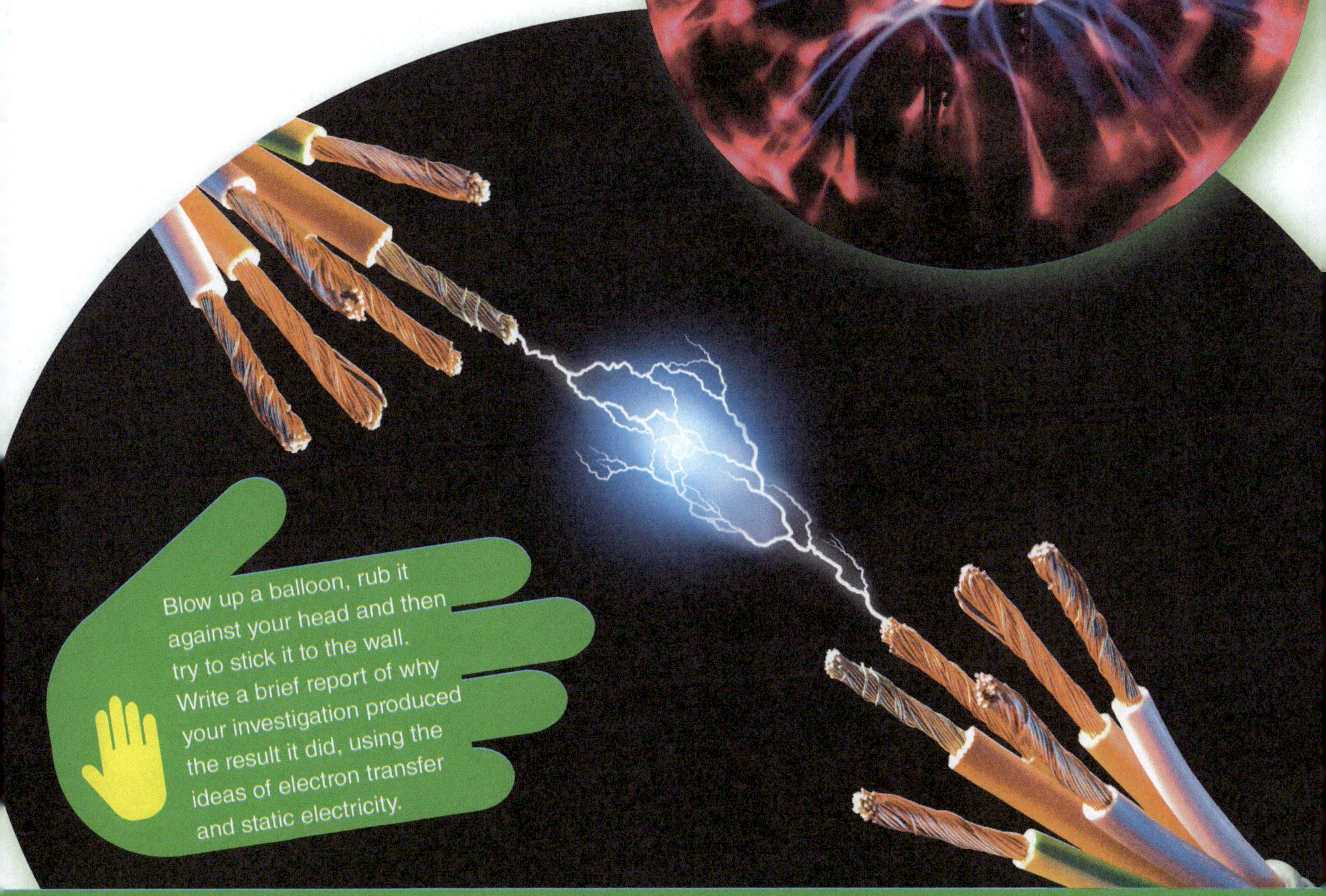

The electric field from an isolated positive charge

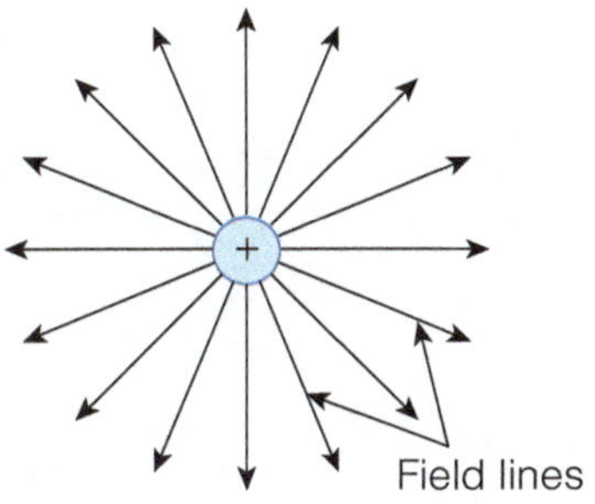

The electric field from an isolated negative charge

Field lines

Blow up a balloon, rub it against your head and then try to stick it to the wall. Write a brief report of why your investigation produced the result it did, using the ideas of electron transfer and static electricity.

1. What causes an object to become positively charged?
2. What happens when two objects with opposing charges are brought together?
3. Why is it important to earth a car when fuelling it?

Mind map

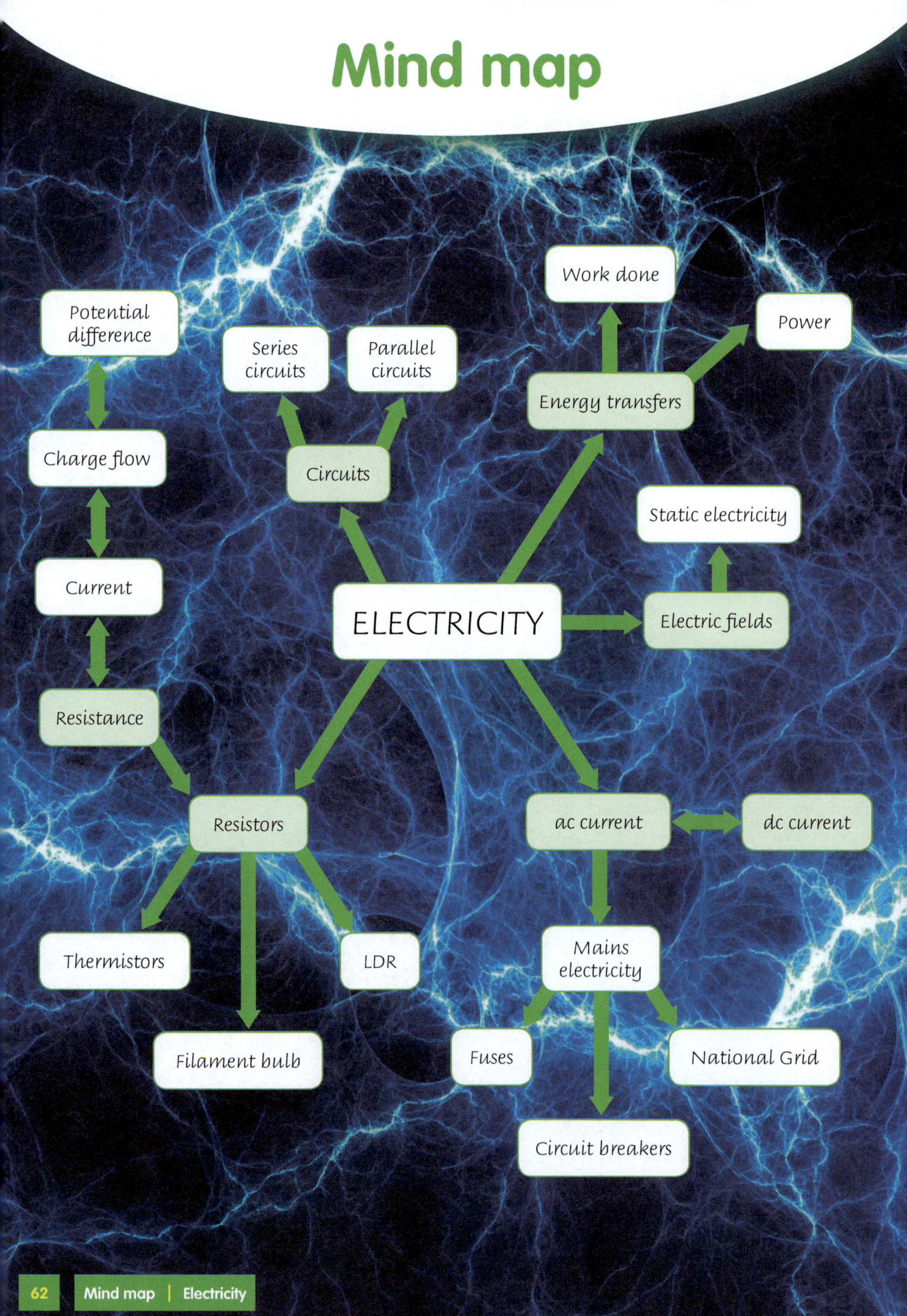

Potential difference

Series circuits

Parallel circuits

Work done

Power

Charge flow

Circuits

Energy transfers

Current

ELECTRICITY

Static electricity

Electric fields

Resistance

Resistors

ac current

dc current

Thermistors

LDR

Mains electricity

Filament bulb

Fuses

National Grid

Circuit breakers

Practice questions

1. Look at the parallel circuit below.

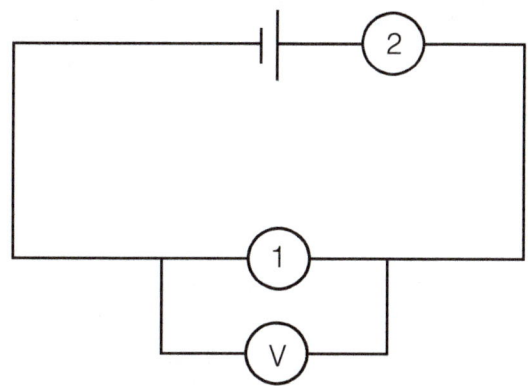

a) Component 1 has a resistance of 3 Ω and has a potential difference of 15 V across it. What is the current through it? **(3 marks)**

b) What is the current at point 2 in the circuit? Explain how you arrived at your answer. **(2 marks)**

c) Sketch a graph to show the relationship between potential difference and current in an ohmic conductor. **(3 marks)**

d) Give an example of a component that would show a non-linear relationship between potential difference and current. **(1 mark)**

2. In an investigation into static electricity, a plastic rod was rubbed with a cloth. The rod was then used to pick up very small bits of paper.

a) After being rubbed with the cloth, the rod became positively charged. What can be deduced from the results of this investigation about the charge on the bits of paper? Explain your answer. **(2 marks)**

b) Use the concept of moving electrons to explain the change in charge of the rod during the investigation. **(2 marks)**

c) Explain what would happen if two identical rods that had been rubbed with the same type of cloth were brought close together. **(2 marks)**

Permanent and induced magnetism, magnetic forces and fields

Poles of a magnet

The poles of a magnet are the places where the magnetic forces are strongest. When two magnets are brought close together they exert a force on each other.

> Two like poles repel. Two unlike poles attract.

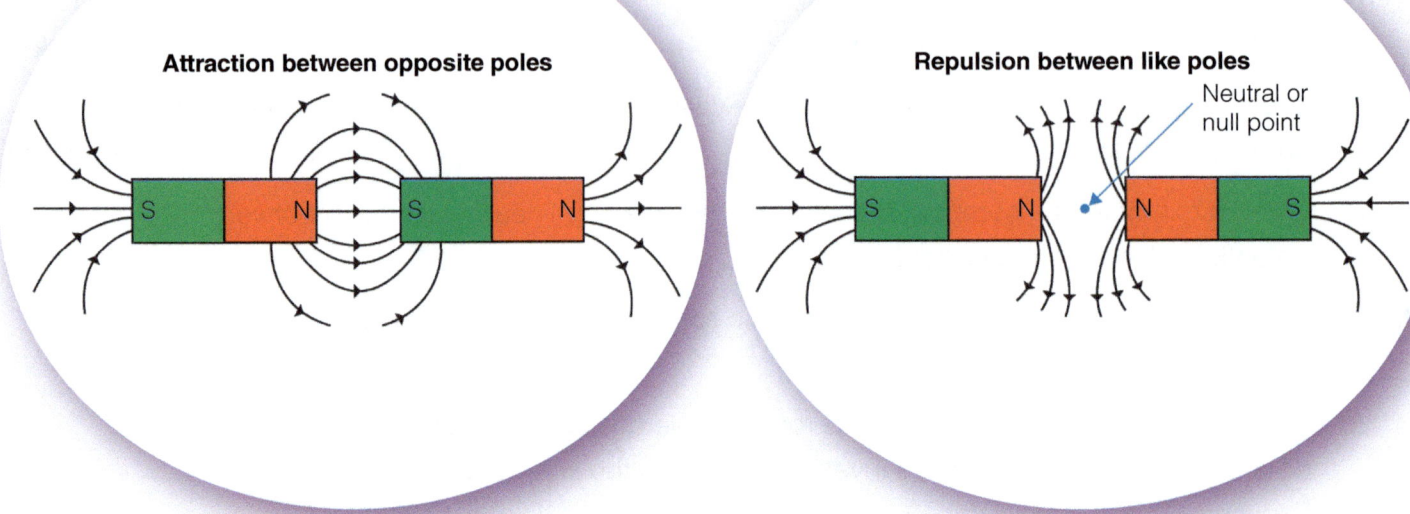

Attraction between opposite poles

Repulsion between like poles

Neutral or null point

Magnetism is an example of a non-contact force.

Keywords

Permanent magnet ➤ Magnet which produces its own magnetic field

Magnetic field ➤ Region around a magnet where a force acts on another magnet or on a magnetic material

Permanent magnetism vs induced magnetism

A **permanent magnet** . . .

➤ produces its own **magnetic field**.

An induced magnet . . .

➤ becomes a magnet when placed in a magnetic field
➤ always experiences a force of attraction
➤ loses most or all of its magnetism quickly when removed from a magnetic field.

Magnetic field

The region around a magnet – where a force acts on another magnet or on a magnetic material (iron, steel, cobalt, magnadur and nickel) – is called the magnetic field.

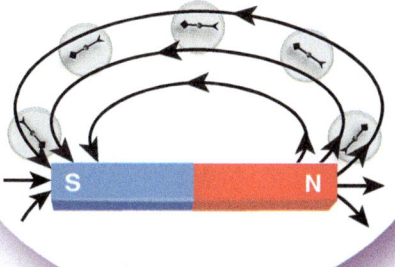

A compass can be used to plot a magnetic field

The **force** between a magnet and a magnetic material is always attraction.

The **strength** of the magnetic field depends on the distance from the magnet.

The **field** is strongest at the poles of the magnet.

The **direction** of a magnetic field line is from the north pole of the magnet to the south pole of the magnet.

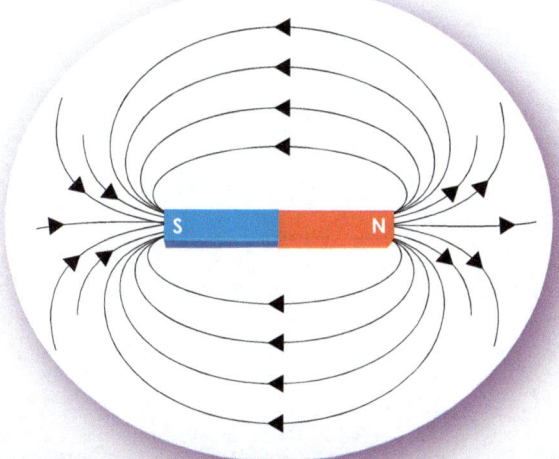

Compasses

A magnetic compass contains a bar magnet that points towards magnetic north. This provides evidence that the Earth's core is magnetic and produces a magnetic field.

26

Make your own compass using a bowl of water, a small dish, a magnet and a nail. To make a magnet for your compass, take the nail and magnet and stroke one end of the magnet along the length of the nail, always going in the same direction. Once your nail is magnetised, you need to float it in the water. Start by putting a dish into the water, then place the nail into the centre of the dish. Let it settle a bit, to move around a little until it has found north.

1. What happens if the two north poles of a bar magnet are brought together?
2. When will a magnetic material become an induced magnet?
3. Where is the magnetic field of a magnet strongest?
4. Why does a magnetic compass point north?

Fleming's left-hand rule, electric motors and loudspeakers

Electromagnets

When a current flows through a conducting wire a magnetic field is produced around the wire.

The shape of the magnetic field can be seen as a series of concentric circles in a plane, perpendicular to the wire.

The direction of these field lines depends on the direction of the current.

The strength of the magnetic field depends on the current through the wire and the distance from the wire.

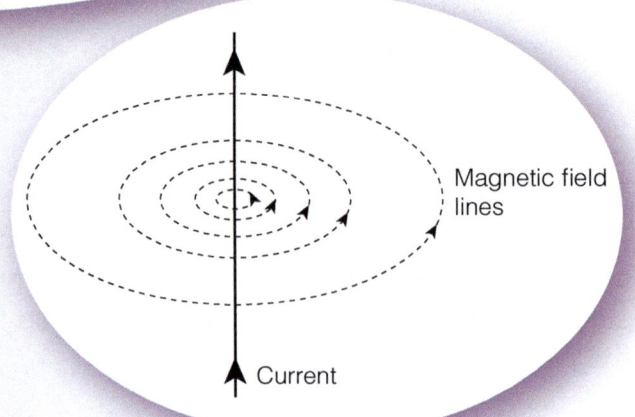

Magnetic field lines

Current

Coiling the wire into a **solenoid** (a helix) increases the strength of the magnetic field created by a current through the wire.

Features of a solenoid

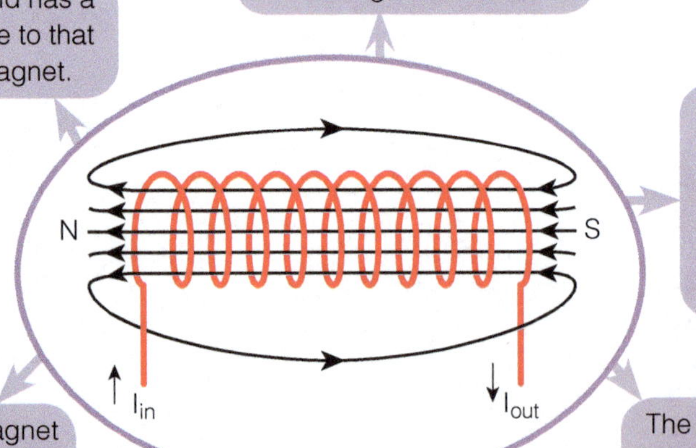

Magnetic field has a similar shape to that of a bar magnet.

Adding an iron core increases the magnetic field strength of a solenoid.

The fields from individual coils in the solenoid add together to form a very strong, almost uniform, field along the centre of the solenoid.

N   S

I_{in} I_{out}

An electromagnet is a solenoid with an iron core.

The fields cancel to give a weaker field outside the solenoid.

27

Electromagnetic coil

Go around the house and make a list of all the devices that contain an electric motor. For each one, explain why the motor is important to the device's function. (Hint: Any device that is electrically powered and has moving parts will have a motor.)

Keyword

Solenoid ➤ Coil wound into a helix shape

Fleming's left-hand rule and the motor effect

When a conductor carrying a current is placed in a magnetic field, the magnet producing the field and the conductor exert an equal and opposite force on each other. This is the motor effect and is due to interactions between magnetic fields.

The direction of the force on the conductor can be identified using Fleming's left-hand rule.

If the direction of the current or the direction of the magnetic field is reversed, the direction of the force on the conductor is reversed.

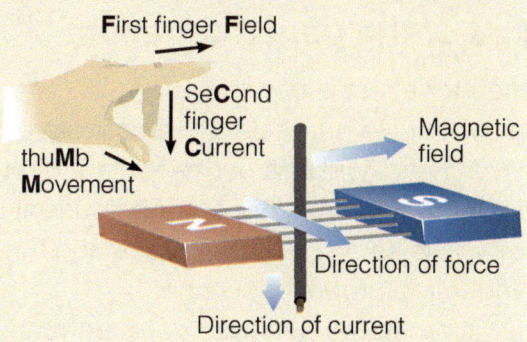

The size of the force on the conductor depends on:

➤ the magnetic flux density
➤ the current in the conductor
➤ the length of conductor in the magnetic field.

For a conductor at right angles to a magnetic field and carrying a current, the force can be calculated using the following equation:

force = magnetic flux density × current × length

$$F = BIl$$

➤ force, F, in newtons, N
➤ magnetic flux density, B, in tesla, T
➤ current, I, in amperes, A (or amp)
➤ length, l, in metres, m

Example:
What is the force produced by a 0.5 m long conductor, with a magnetic flux of 1.2 T and a current of 16 A flowing through it?

$F = BIl$
$F = 1.2 \times 16 \times 0.5 = 9.6$ N

Electric motors

A coil of wire carrying a current in a magnetic field experiences a force, causing it to rotate. This is the basis of an **electric motor**.

The commutator and graphite brush allow the current to be reversed every half turn to keep the coil spinning.

Simple electric motor

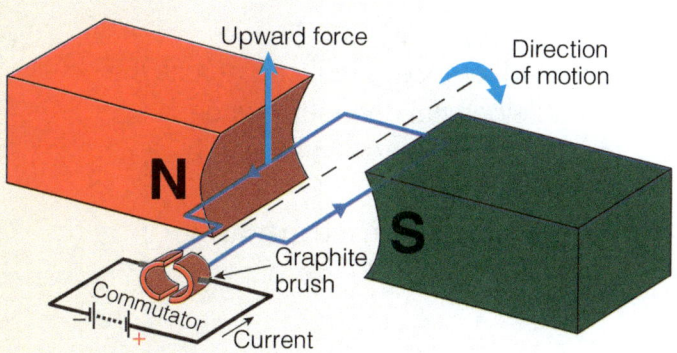

Loudspeakers and headphones

Loudspeakers and headphones use the motor effect to convert variations in current in electrical circuits to the pressure variations in sound waves.

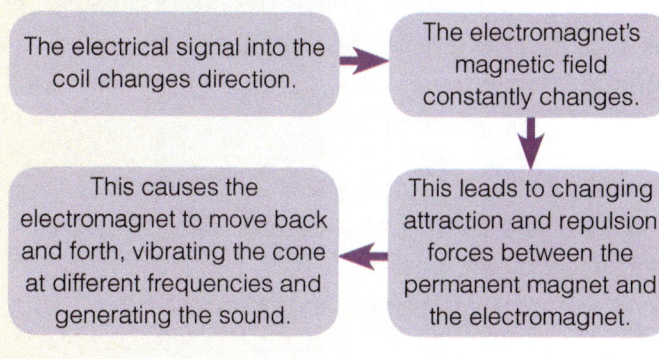

The electrical signal into the coil changes direction.

The electromagnet's magnetic field constantly changes.

This causes the electromagnet to move back and forth, vibrating the cone at different frequencies and generating the sound.

This leads to changing attraction and repulsion forces between the permanent magnet and the electromagnet.

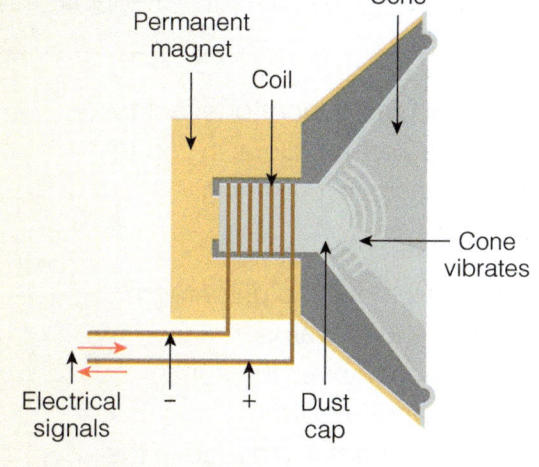

1. What two things does the strength of a magnetic field around a wire depend on?
2. What shape is the magnetic field around a solenoid?
3. What three variables are related by Fleming's left-hand rule?

Induced potential and transformers

HT

Keyword

Transformer ➤ Device used to increase or lower the voltage of an alternating current

28

The generator effect

A potential difference is induced across the ends of an electrical conductor when:

➤ the conductor moves relative to a magnetic field

➤ there is a change in the magnetic field around a conductor.

Induced current

When a potential difference is induced in a conductor that is part of a complete circuit, a current is induced.

The induced current generates a magnetic field. This magnetic field opposes the original change that generated the potential difference – either the movement of the conductor or the change in magnetic field.

The size of the induced potential difference, and therefore the induced current, can be increased by either:

➤ increasing the speed of movement

or

➤ increasing the strength of the magnetic field.

The direction of the induced potential difference and the induced current is reversed if either the direction of movement of the conductor is reversed or the polarity of the magnetic field is reversed.

WS Uses of the generator effect

The generator effect is used:

➤ In a dynamo to generate dc. A coil rotates inside the magnetic field of a permanent magnet, inducing a current in the coil. The current always flows in the same direction, producing the dc voltage-time graph below right.

➤ In an alternator to generate ac. A permanent magnet rotates inside a coil of wire, inducing the current in the coil. The current regularly reverses direction, producing the ac voltage-time graph below left.

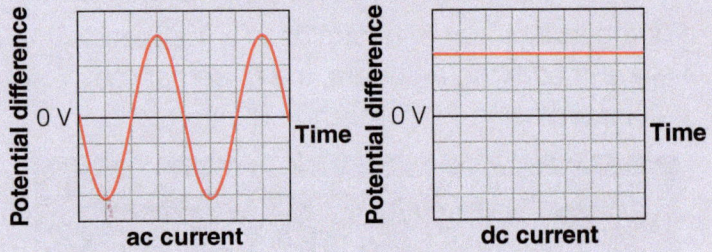

ac current

dc current

Microphones use the generator effect to convert the pressure variations in sound waves into variations in current in electrical circuits.

1 Sound waves move the diaphragm backwards and forwards.

2 The diaphragm moves the coil backwards and forwards.

3 A changing current is induced in the coil that corresponds to the frequency of the sound.

4 This current is used to record the sound or it is fed into an amplifier and into a speaker to produce the sound.

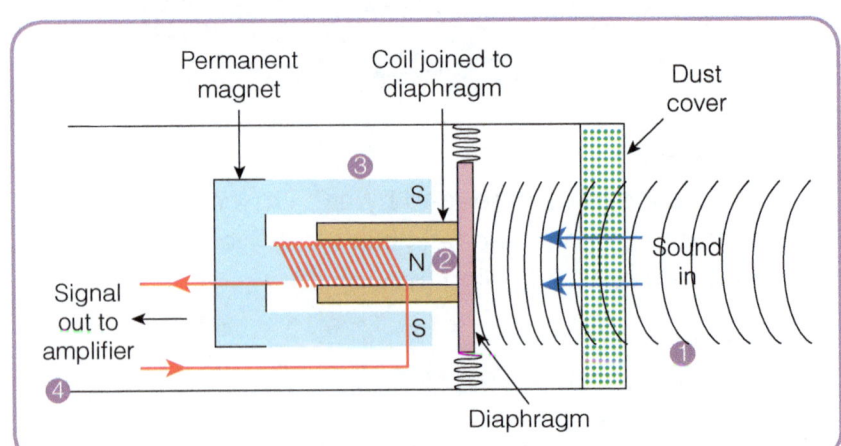

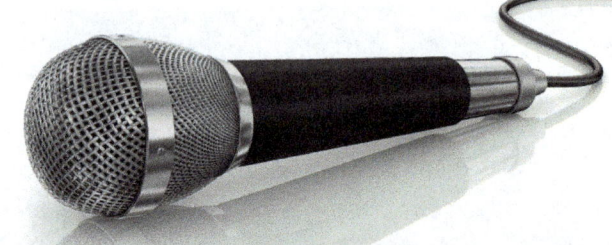

Transformers

A basic **transformer** has a primary coil wire and a secondary coil of wire wound on an iron core. Iron is used as it is easily magnetised.

Primary coil   Iron core   Secondary coil

An alternating current in the primary coil of a transformer produces a changing magnetic field in the iron core.	The changing magnetic field is also produced in the secondary coil.	An alternating potential difference is induced across the ends of the secondary coil.	An induced current will flow in the secondary coil if it is part of a complete circuit.

The relationship between voltage and number of turns of the coil is given by the following equation:

$$\left[\frac{v_P}{v_S} = \frac{n_P}{n_S}\right]$$

➤ **potential difference on primary and secondary coils, v_P and v_S in volts, V**
➤ **number of turns on the primary and secondary coil n_P and n_S**

In a step-up transformer, v_P is greater than v_P. In a step-down transformer, v_S is less than v_P.

The power output of a transformer relates to the power input and the potential difference in the coils.

$V_s \times I_s = V_p \times I_p$
➤ **$V_s \times I_s$ is the power output (secondary coil)**
➤ **$V_p \times I_p$ is the power input (primary coil)**
➤ **power input and output, in watts, W**

Example:

The power output required from a transformer is 500 W. The potential difference on the primary coil is 25 V. What current is required to produce this power output?

$$V_s \times I_s = V_p \times I_p$$
$$V_s \times I_s = 500 \text{ W}$$ ← Required power output = 500 W
$$500 \text{ W} = V_p \times I_p$$
$$500 \text{ W} = 25 \times I_p$$ ← V_p = 25 V
$$\frac{500}{25} = I_p$$
$$20 \text{ A} = I_p$$

The equation linking potential difference and number of turns can be applied together with the equation for power transfer.

Example:

What number of turns on the secondary coil are required to produce a power output of 23 kW, if:

➤ the potential difference in the primary coil is 400 kV
➤ the primary coil has 26 000 turns
➤ the output current is 100 A?

$$V_s \times I_s = 23 \text{ kW}$$
$$100 \times V_s = 23 \text{ kW}$$
$$V_s = \frac{23}{100} = 0.23 \text{ kV}$$
$$\frac{N_p}{N_s} = \frac{400}{0.23} = 1739$$ ← $\left[\frac{v_p}{v_s} = \frac{n_p}{n_s}\right]$
$$\frac{N_p}{N_s} = 1739$$
$$\frac{26000}{N_s} = 1739, \quad \frac{26000}{1739} = N_s$$
$$N_s = 15 \text{ turns}$$ ← this is a step-down transformer

By transmitting the power at high voltage through power cables, the current is kept low, reducing loss of heat energy. A step-down transformer can then be used to lower the voltage to a safer level.

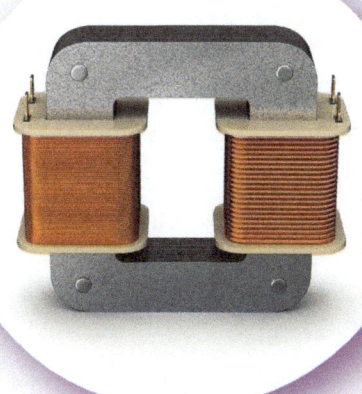

Write down the stages of how a loudspeaker (see Module 27) and a microphone work. Mix them up and rearrange them to show how the sound of a voice is picked up by a microphone and then projected by a speaker.

1. Why is an iron core used in a transformer?
2. What type of current does a dynamo produce?
3. In what two ways can the induced current from a generator be increased?

Mind map

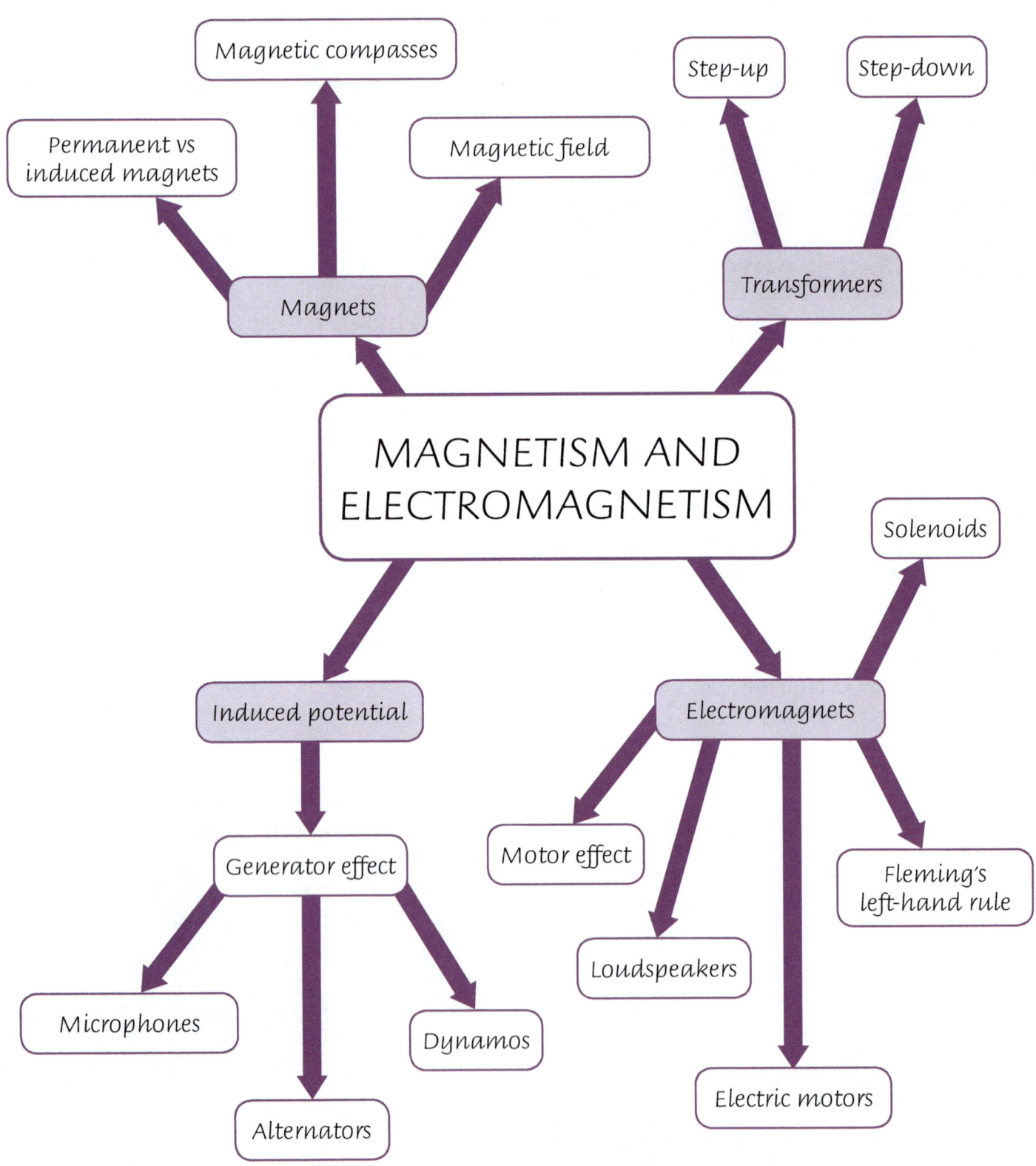

Magnetic compasses

Permanent vs induced magnets

Magnetic field

Magnets

Step-up

Step-down

Transformers

MAGNETISM AND ELECTROMAGNETISM

Solenoids

Induced potential

Electromagnets

Generator effect

Motor effect

Fleming's left-hand rule

Microphones

Dynamos

Loudspeakers

Alternators

Electric motors

Practice questions

1. The diagram below shows a bar magnet.

    a) Draw the shape of the magnetic field around this magnet.   **(2 marks)**

    b) What would happen if the south pole of a bar magnet was brought into contact
        with the north pole of another magnet?   **(1 mark)**

    c) Give two differences between a bar magnet and an induced magnet.   **(2 marks)**

2. The diagram shows a transformer.

    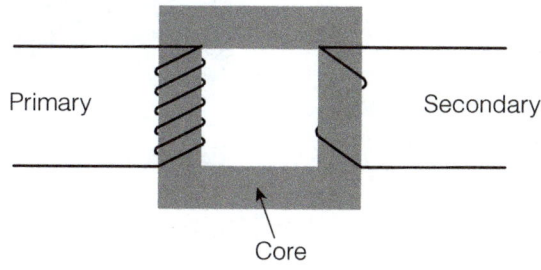

    Is this a step-up transformer or a step-down transformer?
    How do you know?   **(3 marks)**

3. Explain the importance of transformers to the National Grid.   **(3 marks)**

4. If the transformer shown in the diagram had a potential difference of 15 V in its
    secondary coil, what would be the voltage in its primary coil?   **(2 marks)**

Changes of state and the particle model

The particle model

Matter can exist as a solid, liquid or as a gas.

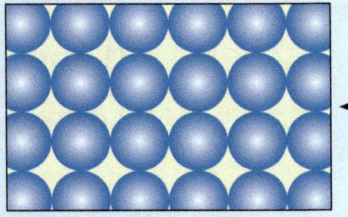

Solid – particles are very close together and vibrating. They are in fixed positions.

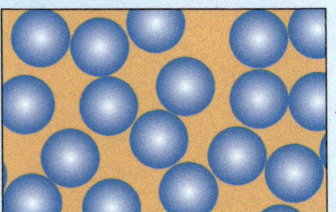

Liquid – particles are very close together but are free to move relative to each other. This allows liquids to flow.

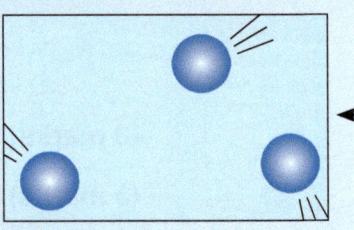

Gas – particles in a gas are not close together. The particles move rapidly in all directions.

If the particles in a substance are more closely packed together, the density of the substance is higher. This means that liquids have a higher density than gases. Most solids have a higher density than liquids.

Density also increases when the particles are forced into a smaller volume.

Low density

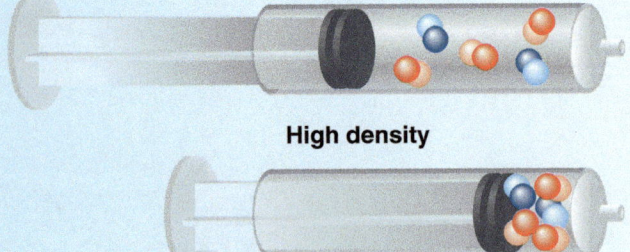

High density

The density of a material is defined by the following equation:

$$density = \frac{mass}{volume}$$

$$\rho = \frac{m}{V}$$

➤ density, ρ, in kilograms per metre cubed, kg/m³
➤ mass, m, in kilograms, kg
➤ volume, V, in metres cubed, m³

Example:
What is the density of an object that has a mass of 56 kg and a volume of 0.5 m³?

$$\rho = \frac{m}{V}$$

$$= \frac{56}{0.5} = 112 \text{ kg/m}^3$$

When substances change state (melt, freeze, boil, evaporate, condense or sublimate), mass is conserved (it stays the same).

Changes of state are physical changes: the change does not produce a new substance, so if the change is reversed the substance recovers its original properties.

Ice	Water	Steam
		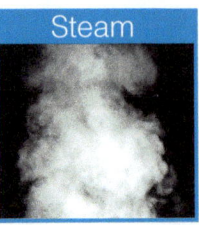

29

Internal energy

Energy is stored inside a system by the particles (atoms and molecules) that make up the system. This is called **internal energy**.

Internal energy of a system is equal to the total kinetic energy and potential energy of all the atoms and molecules that make up the system.

Heating changes the energy stored within the system by increasing the energy of the particles that make up the system. This either raises the temperature of the system or produces a change of state.

Heat and temperature are related but are not a measure of the same thing.

➤ Heat is the amount of thermal energy and is measured in Joules (J).
➤ Temperature is how hot or cold something is and is measured in degrees Celsius (°C).

Changes of heat and specific latent heat

When a change of state occurs, the stored internal energy changes, but the temperature remains constant. The graph below shows the change in temperature of water as it is heated; the temperature is constant when the water is changing state.

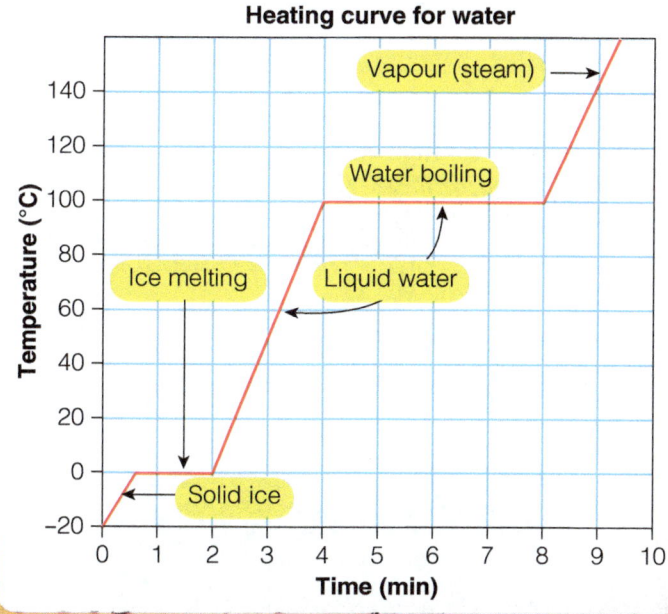

Heating curve for water

Keyword

Specific latent heat ➤ The energy required to change the state of one kilogram of the substance with no change in temperature

The **specific latent heat** of a substance is equal to the energy required to change the state of one kilogram of the substance with no change in temperature.

The energy required to cause a change of state can be calculated by the following equation:

energy for a change of state = mass × specific latent heat

$$E = mL$$

➤ energy, E, in joules , J
➤ mass, m, in kilograms, kg
➤ specific latent heat, L, in joules per kilogram, J/kg

Example:
What is the energy needed for 600 g of water to melt? (The specific latent heat of water melting is 334 kJ/kg.)

$$E = mL$$
$$0.6 \times 334 = 200.4 \text{ kJ}$$

The specific latent heat of fusion is the energy required for a change of state from solid to liquid.

The specific latent heat of vapourisation is the energy required for a change of state from liquid to vapour

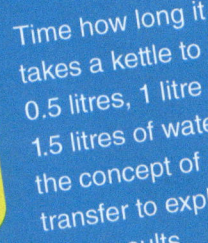

Time how long it takes a kettle to boil 0.5 litres, 1 litre and 1.5 litres of water. Use the concept of energy transfer to explain your results.

1. What energy is needed for 200 kg of iron to turn from a solid to a liquid? (The specific latent heat of iron melting is 126 kJ/kg.)
2. Explain the difference between the specific latent heat of fusion and the specific latent heat of vapourisation.
3. If a substance is heated but its temperature remains constant, what is happening to the substance?

Particle model and pressure

Keyword
Gas pressure ➤ Total force exerted by all of the gas molecules inside the container on a unit area of the wall

Gas under pressure

The molecules of a gas are in constant random motion.

When the molecules collide with the wall of their container they exert a force on the wall. The total force exerted by all of the molecules inside the container on a unit area of the wall is the **gas pressure**.

Increasing the temperature of a gas, held at constant volume, **increases** the pressure exerted by the gas.

Decreasing the temperature of a gas, held at constant volume, **decreases** the pressure exerted by the gas.

The temperature of the gas is related to the average kinetic energy of the molecules. The higher the temperature, the greater the average kinetic energy, and so the faster the average speed of the molecules. At higher temperatures, the particles collide with the walls of the container at a higher speed.

30

Draw a comic strip showing the effect of changing volume and temperature on a gas. Annotate your comic strip using the terms and equations in this module.

Temperature can be measured in degrees Celsius (°C) or Kelvin (K). To convert from Celsius to Kelvin, add 273.

For example:
➤ 10°C + 273 = 283 K

0 Kelvin (−273°C) is **absolute zero**. At this point, the particles have no kinetic energy so are not moving.

A gas can be compressed or expanded by pressure changes. The pressure produces a net force at right angles to the wall of the gas container (or any surface).

When the volume of the gas in the container is reduced, the pressure increases.

High pressure – Low volume

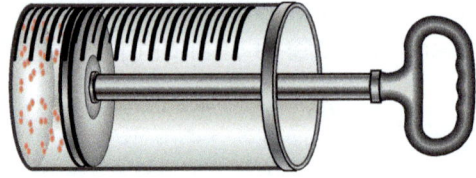

Low pressure – High volume

For a fixed mass of gas held at a constant temperature this equation is used:

constant = pressure × volume
constant = pV
➤ pressure, p, in pascals, Pa
➤ volume, V, in metres cubed, m^3

When a volume of a gas is altered, this allows the new pressure to be calculated by the following equation:

$$p_1 \times V_1 = p_2 \times V_2$$
➤ p_1 and V_1 are the initial pressure and volume
➤ p_2 and V_2 are the final pressure and volume

Example:
A gas with a pressure of 200 kPa is compressed from a volume of 3 m³ to a volume of 0.5 m³. What is the new pressure of the gas?

$$p_1 \times V_1 = p_2 \times V_2$$
$$200 \times 3 = p_2 \times 0.5$$
$$600 = p_2 \times 0.5$$
$$\frac{600}{0.5} = p_2$$
$$1200 \text{ kPa} = p_2$$

 Doing work on a gas increases the internal energy of the gas and can cause an increase in the temperature of the gas. Look at this flow chart:

Pumping the handle of a bicycle pump is work done.

Energy is transferred to the air particles in the pump.

This increases the internal energy of the air particles.

As the particles in the air have more internal energy, the temperature of the air in the pump increases.

1. Explain how gases exert pressure.
2. What is 32°C in Kelvin?
3. When work is done to a gas, why does the temperature increase?

Mind map

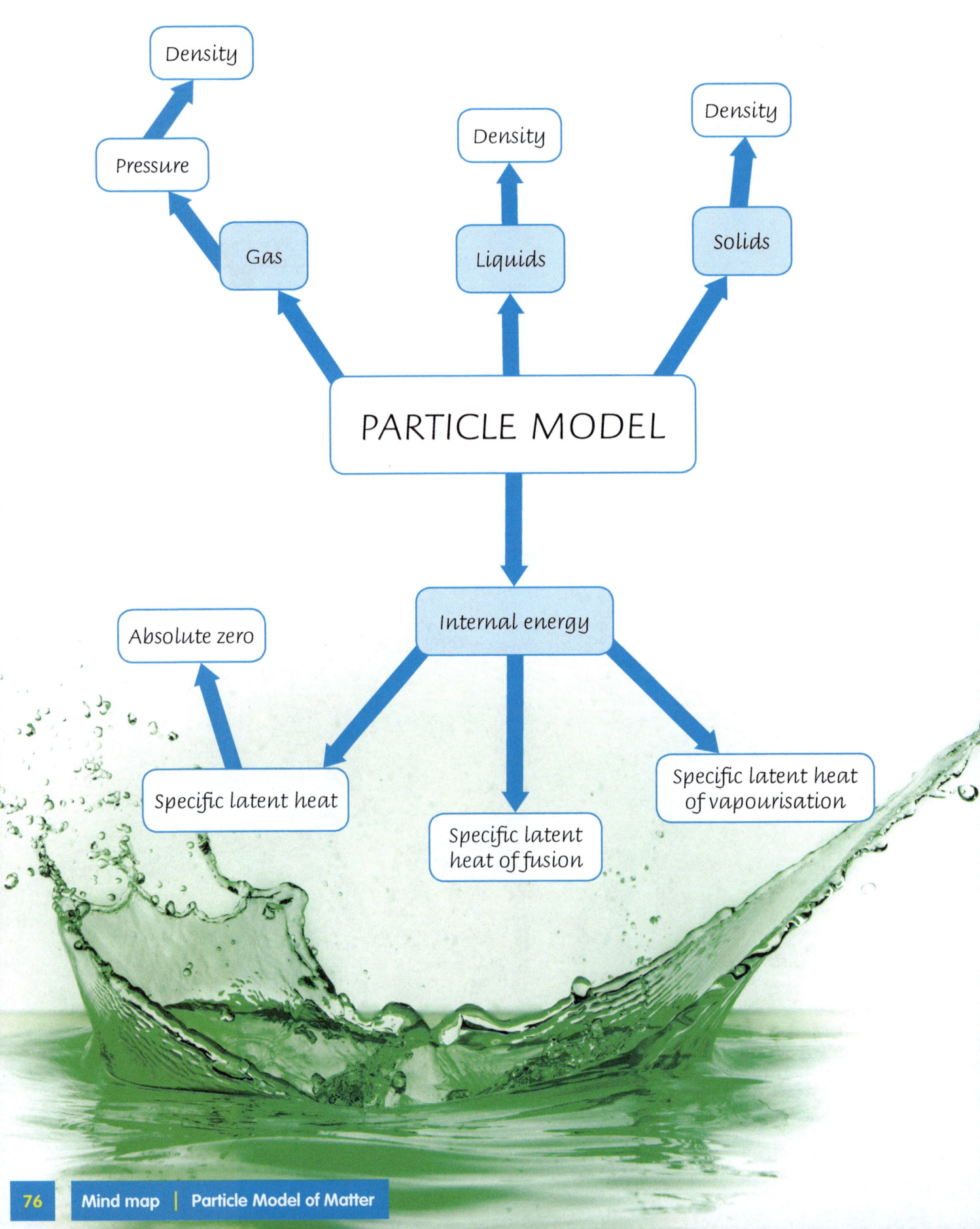

Density

Pressure

Gas

Density

Liquids

Density

Solids

PARTICLE MODEL

Internal energy

Absolute zero

Specific latent heat

Specific latent heat of fusion

Specific latent heat of vapourisation

Practice questions

1. A solid has a mass of 150 g and a volume of 0.0001 m³.

    a) What is the density of the solid? **(2 marks)**

    b) What happens to the mass and density of the solid when it sublimes? **(2 marks)**

    c) Explain what would happen to the stored internal energy and the temperature of the solid as it sublimed. **(2 marks)**

    d) The specific latent heat of the solid is 574 kJ/kg.

    What is the energy required for the solid to sublime? **(2 marks)**

2. a) Explain, using the particle model, why:

    i) a gas at a constant volume has a higher pressure when the temperature increases. **(2 marks)**

    ii) a gas at a constant temperature has a lower pressure when the volume increases. **(2 marks)**

    b) What is the new volume of a gas that changes its pressure from 500 kPa to 134 kPa? The initial volume was 2.1 m³. **(3 marks)**

3. A solid has a temperature of 300 Kelvin.

    a) What is this temperature in °C? **(1 mark)**

    b) If the temperature fell to 250 Kelvin, how would the average kinetic energy and average speed of the particles in the solid change? **(1 mark)**

4. What term is given to the energy required for 1 kg of a liquid to become a gas? **(1 mark)**

Atoms and isotopes

Structure of atoms

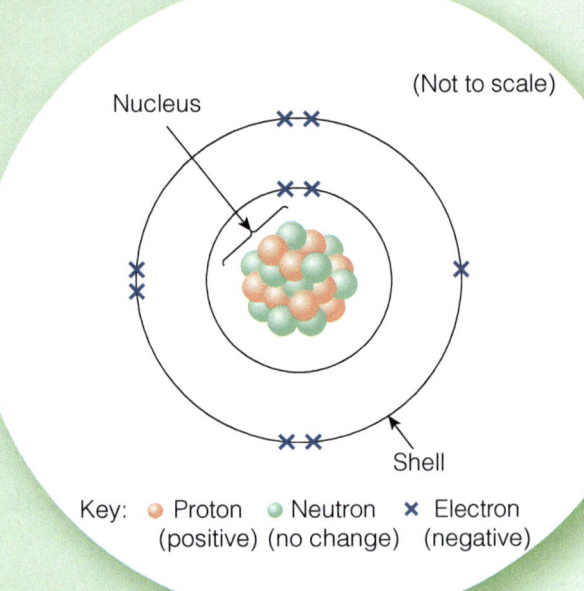

Nucleus

(Not to scale)

Shell

Key: ● Proton (positive) ● Neutron (no change) ✕ Electron (negative)

Atoms have a radius of around 1×10^{-10} metres.

The radius of a nucleus is less than $\dfrac{1}{10\ 000}$ of the radius of an atom.

Most of the mass of an atom is concentrated in the nucleus. Protons and neutrons have a relative mass of 1 while electrons have a relative mass of 0.0005. The electrons are arranged at different distances from the nucleus (are at different energy levels).

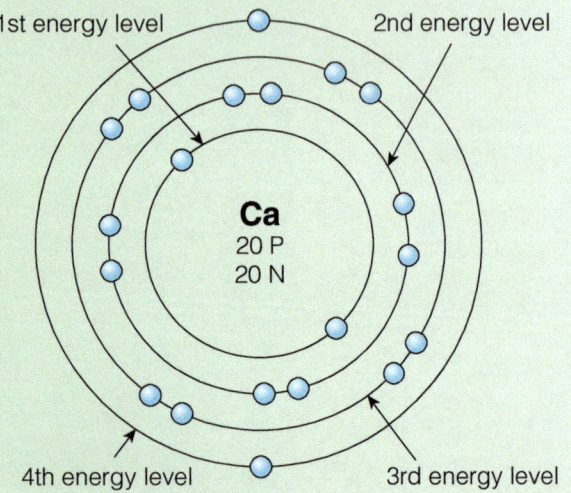

1st energy level   2nd energy level

Ca
20 P
20 N

4th energy level   3rd energy level

Absorption of electromagnetic radiation causes the electrons to become excited and move to a higher energy level and further from the nucleus.

Emission of electromagnetic radiation causes the electrons to move to a lower energy level and move closer to the nucleus.

If an atom loses or gains an electron, it is ionised.

The number of electrons is equal to the number of protons in the nucleus of an atom.

Atoms have no overall electrical charge.

All atoms of a particular element have the same number of protons. The number of protons in an atom of an element is called the **atomic number**.

The total number of protons and neutrons in an atom is called the **mass number**.

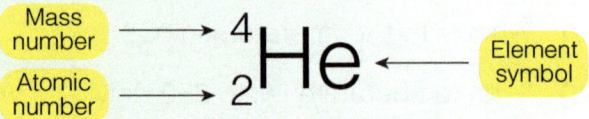

Mass number ⟶ ^{4}
Atomic number ⟶ $_{2}$He ⟵ Element symbol

Atoms of the same element can have different numbers of neutrons; these atoms are called isotopes of that element. For example, below are some isotopes of nitrogen. They each have 7 protons in the nucleus but different numbers of neutrons, giving the different isotopes.

$$^{14}N \quad ^{15}N \quad ^{13}N$$

Atoms turn into **positive ions** if they lose one or more outer electrons and into **negative ions** if they gain one or more outer electrons.

Write out the stages of the discovery of the atom on small cards. Mix them up then put them in the correct order.

The development of the atomic model

Keywords
Atomic number ➤ Number of protons in an atom
Mass number ➤ Total number of protons and neutrons in an atom

 Before the discovery of the electron, atoms were thought to be tiny spheres that could not be divided.

The discovery of the electron led to further developments of the model. The plum pudding model suggested that the atom is a ball of positive charge with negative electrons embedded in it.

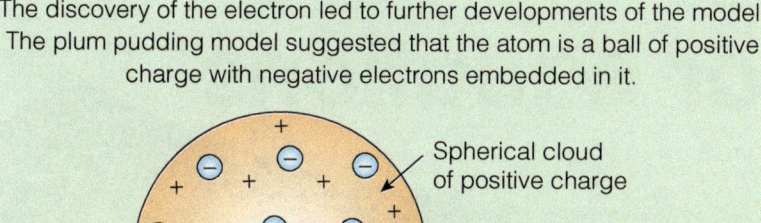

Spherical cloud of positive charge

Electron

Rutherford, Geiger and Marsden's alpha scattering experiment led to the conclusion that the mass of an atom was concentrated at the centre (nucleus) and that the nucleus was charged.

The nucleus

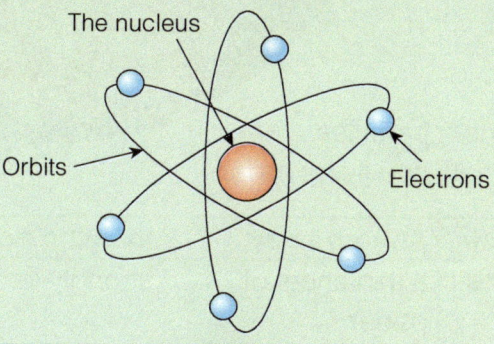

Orbits

Electrons

This evidence led to the nuclear model replacing the plum pudding model.

Niels Bohr suggested that the electrons orbit the nucleus at specific distances. The theoretical calculations of Bohr agreed with experimental observation.

Orbital electrons, negatively charged

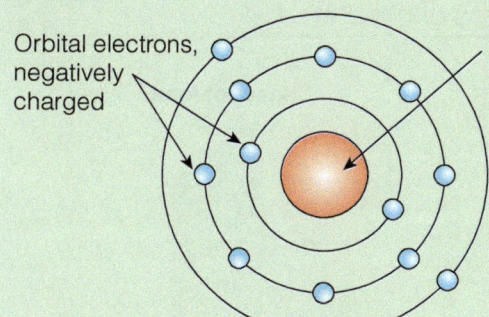

Nucleus, containing positively charged protons

Later experiments led to the idea of the nucleus containing smaller particles with the same amount of positive charge (**protons**).

In 1932, the experimental work of James Chadwick provided evidence of the existence within the nucleus of the **neutron**.

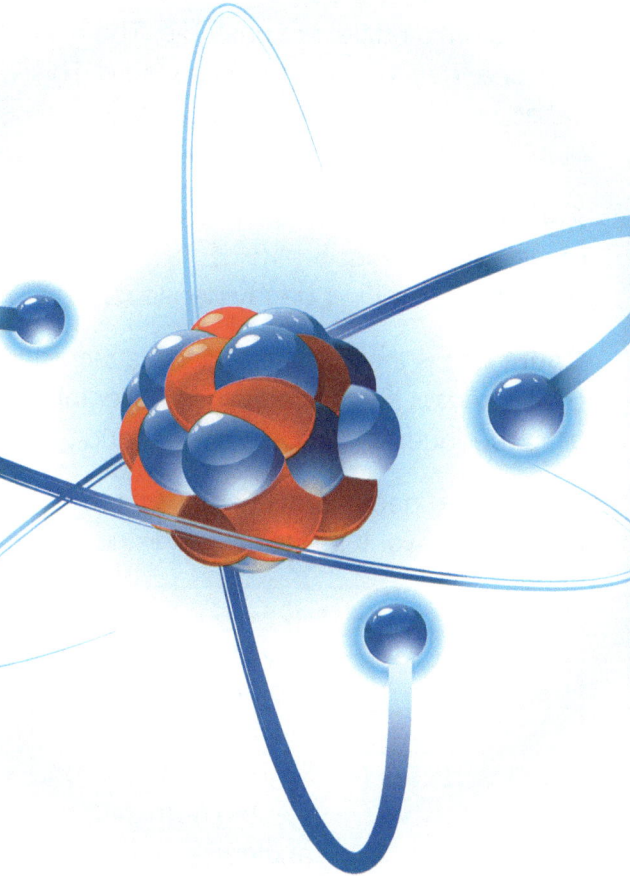

1. What is the effect on the electrons when an atom absorbs electromagnetic radiation?
2. What name is given to atoms of the same element that have different numbers of neutrons?
3. What is the difference between the plum pudding model and the nuclear model?

Radioactive decay, nuclear radiation and nuclear equations

Radioactive decay

Some atomic nuclei are unstable. The nucleus gives out radiation as it changes to become more stable. This is a random process called **radioactive decay**. Changes in atoms and nuclei can also generate and absorb radiation over the whole frequency range.

Activity is the rate at which a source of unstable nuclei decays, measured in **becquerel** (Bq).

➤ 1 becquerel = 1 decay per second

Count rate is the number of decays recorded each second by a detector, such as a Geiger-Müller tube).

➤ 1 becquerel = 1 count per second

Radioactive decay can release a neutron, alpha particles, beta particles or gamma rays. If radiation is ionising, it can damage materials and living cells.

Particle	Description	Penetration in air	Absorbed by...	Ionising power
Alpha particles (α)	Two neutrons and two protons (a helium nucleus).	a few centimetres	a thin sheet of paper	strongly ionising
Beta particles (β)	High speed electron ejected from the nucleus as a neutron turns into a proton.	a few metres	a sheet of aluminium about 5 mm thick	moderately ionising
Gamma rays (γ)	Electromagnetic radiation from the nucleus.	a large distance	a thick sheet of lead or several metres of concrete	weakly ionising

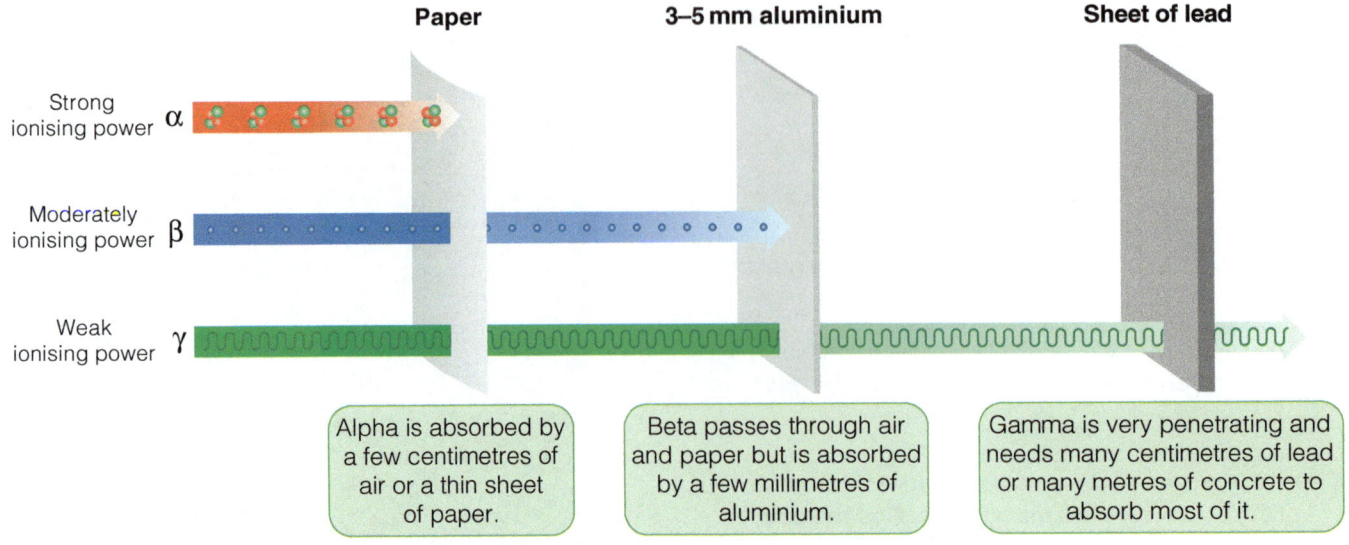

Alpha is absorbed by a few centimetres of air or a thin sheet of paper.

Beta passes through air and paper but is absorbed by a few millimetres of aluminium.

Gamma is very penetrating and needs many centimetres of lead or many metres of concrete to absorb most of it.

Nuclear equations

Nuclear equations are used to represent radioactive decay.

Nuclear equations can use the following symbols:

$^{4}_{2}\text{He}$ alpha particle

$^{0}_{-1}\text{e}$ beta particle

Alpha decay causes both the mass and charge of the nucleus to decrease, as two protons and two neutrons are released.

$$^{219}_{86}\text{radon} \rightarrow {}^{215}_{84}\text{polonium} + {}^{4}_{2}\text{He}$$

Beta decay does not cause the mass of the nucleus to change but does cause the charge of the nucleus to increase, as a proton becomes a neutron.

$$^{14}_{6}\text{carbon} \rightarrow {}^{14}_{7}\text{nitrogen} + {}^{0}_{-1}\text{e}$$

The above example is β– decay as a neutron has becomes a proton and an electron has been ejected. In β+ decay a proton becomes a neutron plus a positron.

The emission of a gamma ray does not cause the mass or the charge of the nucleus to change.

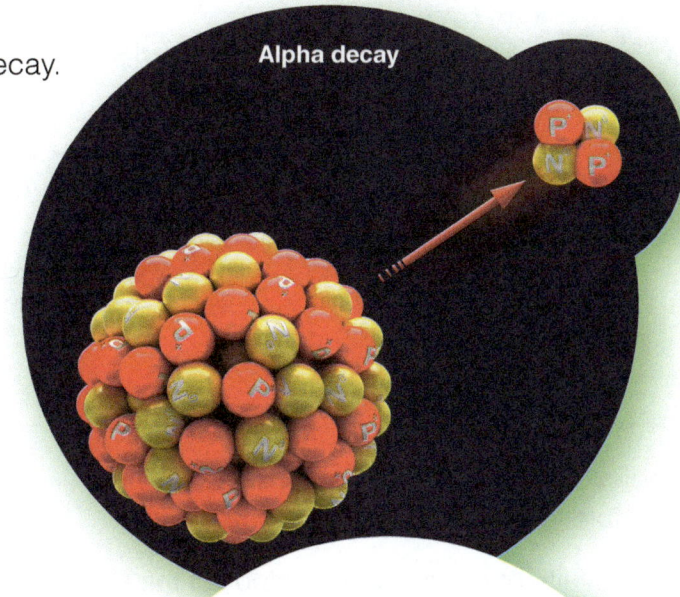

Alpha decay

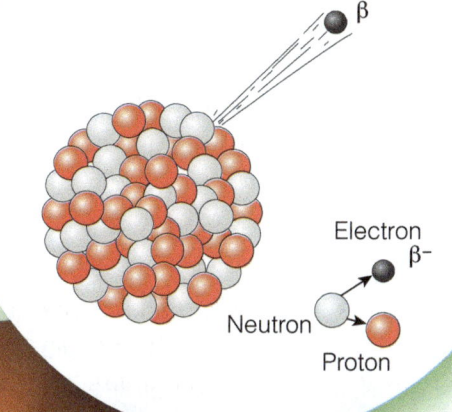

Beta-minus decay with gamma ray

β

Electron β⁻

Neutron

Proton

Make a large version of the radioactive decay table p. 80. Cut out each individual box to form separate cards. Mix the cards up and then group them to show the features of alpha, beta and gamma particles.

Keywords

Radioactive decay ➤ Random release of radiation from an unstable nucleus as it becomes more stable

Becquerel ➤ Unit of rate of radioactive decay

1. How far does alpha radiation penetrate in air?
2. What material is required to absorb gamma rays?
3. What effect does beta decay have on the mass and charge of the nucleus of an atom?

32

Half-lives and the random nature of radioactive decay

Uranium

Half-life

Radioactive decay occurs randomly. It is not possible to predict which nuclei will decay.

The half-life of a radioactive isotope is the average time it takes for:

➤ the number of nuclei in a sample of the isotope to halve

or

➤ the count rate (or activity) from a sample containing the isotope to fall to half of its initial level.

For example:

A radioactive sample has an activity of 560 counts per second. After 8 days, the activity is 280 counts per second. This gives a half-life of 8 days.

The decay can be plotted on a graph and the half-life determined from the graph.

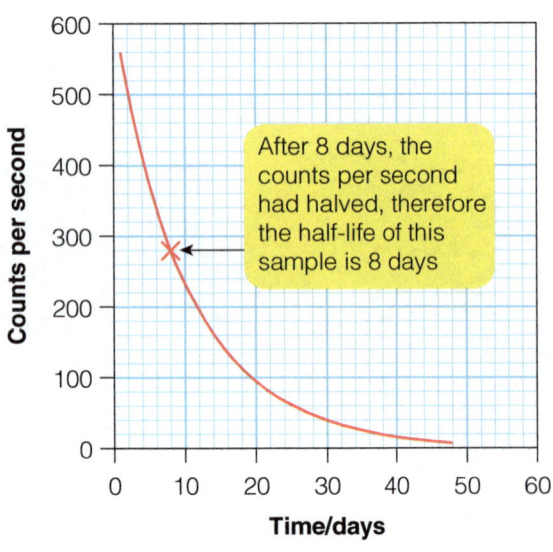

After 8 days, the counts per second had halved, therefore the half-life of this sample is 8 days

HT After another 8 days, the activity would now be 140 counts per second. As a ratio the total net decline is 140 : 560 or 1 : 4.

Radioactive contamination

Radioactive contamination is the unwanted presence of materials containing radioactive atoms or other materials.

This is a hazard due to the decay of the contaminating atoms. The level of the hazard depends on the type of radiation emitted.

Irradiation is the process of exposing an object to nuclear radiation. This is different from radioactive contamination as the irradiated object does not become radioactive.

Suitable precautions must be taken to protect against any hazard from the radioactive source used in the process of irradiation. In medical testing using radioactive sources, the doses patients receive are limited and medical staff wear protective equipment.

Key

Irradiation ➤ Exposing object to nuclear radiation without the object becoming radioactive itself

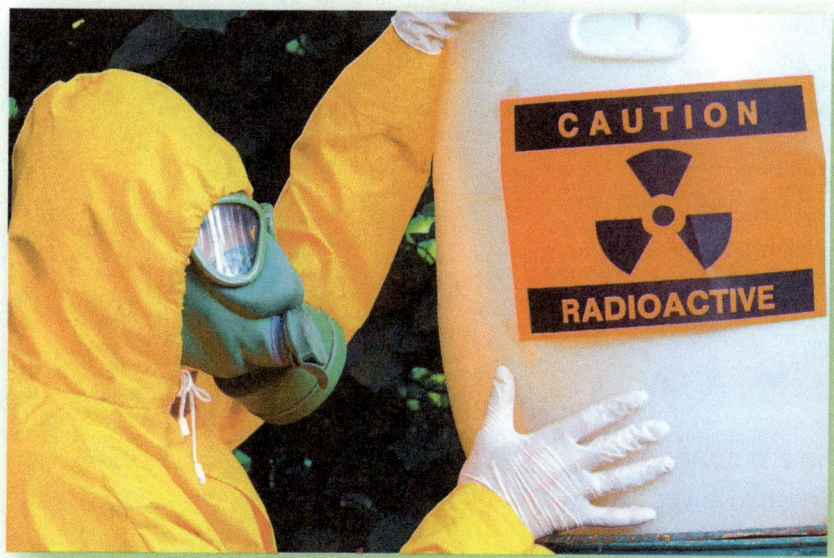

It is important for the findings of studies into the effects of radiation on humans to be published and shared with other scientists. This allows the findings of the studies to be checked by other scientists by the peer review process.

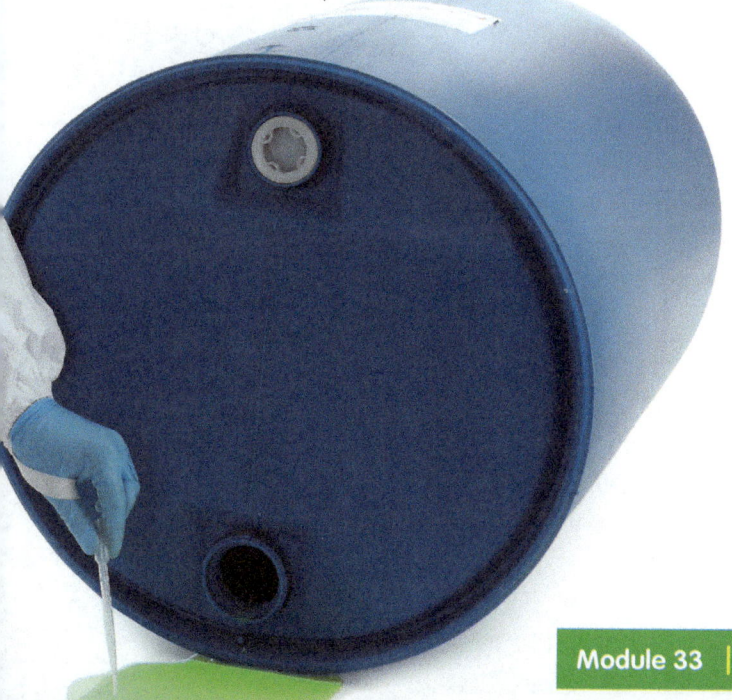

Shake ten or more coins in your hand and then drop them. Remove all the coins that are heads and record the number you've removed. Repeat the experiment until you have one or no coins left. Now plot a graph of how the number of coins in your hand decreased over time. Then, research graphs showing radioactive decay and compare them to the graph from your experiment.

1. Give the two definitions of half-life.
2. What is the half-life of a sample that goes from a count per second of 960 to 240 in 10 months?
3. Why is radioactive contamination a hazard?

Background radiation

Background radiation is around us all the time.

Natural radiation sources	Man-made sources
Some rocks, such as granite.	Fallout from nuclear weapons testing.
Cosmic rays from space.	Fallout from nuclear accidents.

The level of background radiation and the radiation dose received by a person may be affected by their occupation and their location.

Radiation dose is measured in sieverts (Sv).
1000 millisieverts (mSv) = 1 sievert (Sv)

Radioactivity can be detected and measured in a number of ways, including using photographic film or a Geiger-Müller tube.

Cosmic Rays

Granite

Different half-lives of radioactive isotopes

Radioactive isotopes have a wide range of different half-life values.

Shortest half-lives	Longest half-lives
Unstable source	Stable source
Rapid decay	Slow decay
Large amount of radiation emitted in a short time	Small amount of radiation emitted per second but continues to emit radiation over a very long period of time
Example: lithium-4 has a half-life of 324×10^{-24}	Example: calcium-48 has a half-life of 43×10^{18} years

Radioactive isotopes with a long half-life can provide a long-term hazard because they continue to emit small amounts of radiation over a long period of time. Radioactive isotopes with a short half-life can provide a short-term danger as large amounts of radiation are released in a short time.

Keyword

Background radiation
➤ Radiation which is around us all the time

Uses of nuclear radiation

Radioactivity is used in:

- smoke alarms
- irradiating food to improve its safety and extend shelf life
- sterilisation of equipment, such as medical equipment
- tracing and gauging the thickness of a material, e.g. to monitor the structure of pipes.

Nuclear radiations are used in medicine for:

- exploration of internal organs, such as the use of ^{13}N as a radioactive tracer
- control or destruction of unwanted tissue, e.g. gamma rays are used to destroy cancerous tumours.

A radiation source used as a tracer needs to have a long enough half-life to ensure it emits throughout the test period, but short enough that it doesn't continue to emit after the test period is over.

The radiation emitted should also be able to pass through body tissues so it can be detected but not cause any negative health effects for the patient or medical personnel.

As ionising radiation causes tissue damage, it is important to make sure that it is very well targeted when used to destroy cancer cells. Ionising radiation can also lead to mutations so it's important to balance this risk against the risk posed by the tumour to be destroyed.

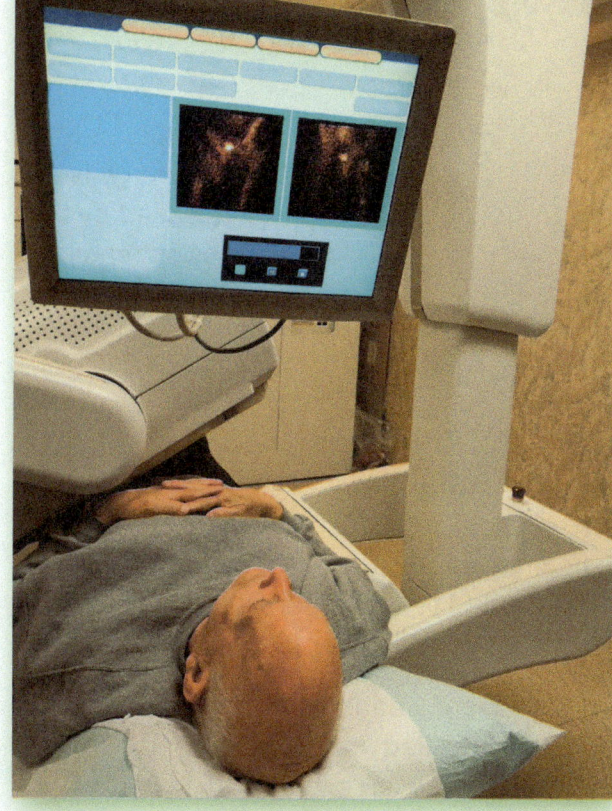

Research the different uses of radiation in medicine and make a poster explaining the differences between the radiation used in tracers and the radiation used to kill cells.

1. Give one man-made source of background radiation and one natural source.
2. Is a highly unstable radioactive source likely to have a long or short half-life?
3. Give one use of nuclear radiation in medicine.

Nuclear fission and fusion

Keywords
Nuclear fission ➤ Splitting of a large and unstable atomic nucleus in a radioactive element
Nuclear fusion ➤ Joining of two light nuclei to form a heavier nucleus

Nuclear fission

Nuclear fission is the splitting of a large and unstable atomic nucleus in a radioactive element, such as uranium or plutonium. Spontaneous fission is rare.

Usually, for fission to occur the unstable nucleus must first absorb a neutron.

Some of the mass of the smaller nuclei may be converted into the energy of radiation.

The nucleus emits two or three neutrons plus gamma rays.

Energy is released by the fission reaction.

The neutrons have kinetic energy and may go on to start a chain reaction.

Chain reaction

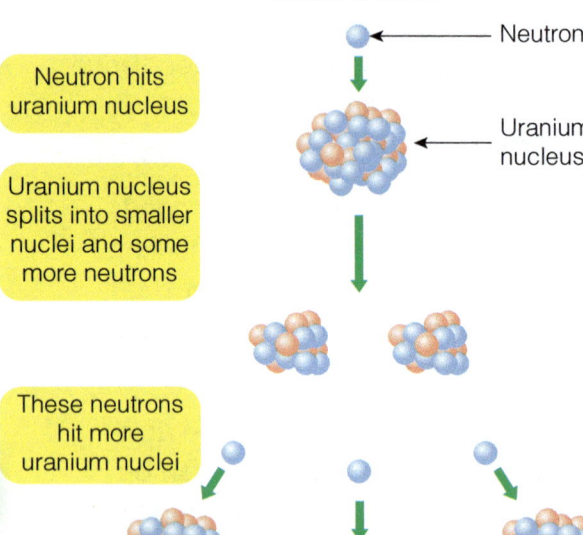

Neutron

Neutron hits uranium nucleus

Uranium nucleus

Uranium nucleus splits into smaller nuclei and some more neutrons

These neutrons hit more uranium nuclei

In a nuclear reactor, the chain reaction is **controlled** to ensure the energy released can be controlled.

In a nuclear weapon, the chain reaction is **uncontrolled**, causing the energy to be released in an explosion.

Nuclear fusion

Nuclear fusion is the joining of two light nuclei to form a heavier nucleus.

In this process some of the mass of the smaller nuclei is converted into energy.

Some of this energy may be the energy of emitted radiation.

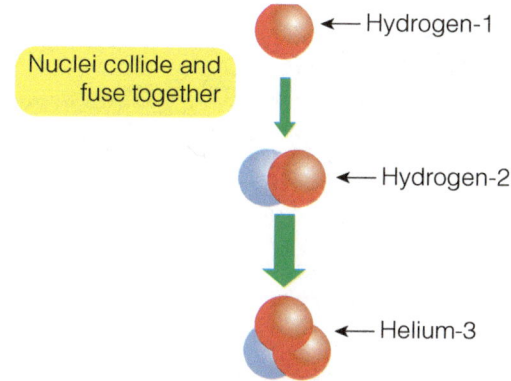

Nuclei collide and fuse together

← Hydrogen-1

← Hydrogen-2

← Helium-3

The protons in the nucleus give it a positive charge. This means that when nuclei are brought close to each other, they repel (electrostatic repulsion).

Very high temperatures and pressures are required to bring the nuclei close enough together and overcome the electrostatic repulsion for fusion to happen. This means that it has not been possible to create a fusion reactor that could be used to generate electricity economically. At the moment, fusion reactors generate less energy than is used to cause the fusion.

Nuclear fusion occurs in stars.

Write down each stage of nuclear fission and fusion on separate cards. Mix the cards up and then rearrange them to form the correct order for fission and fusion.

1. What may the neutrons released by nuclear fission cause?
2. Explain the difference between the release of energy in a nuclear reactor and from a nuclear weapon.
3. Why are very high temperatures and pressures required to cause fusion to happen?

Mind map

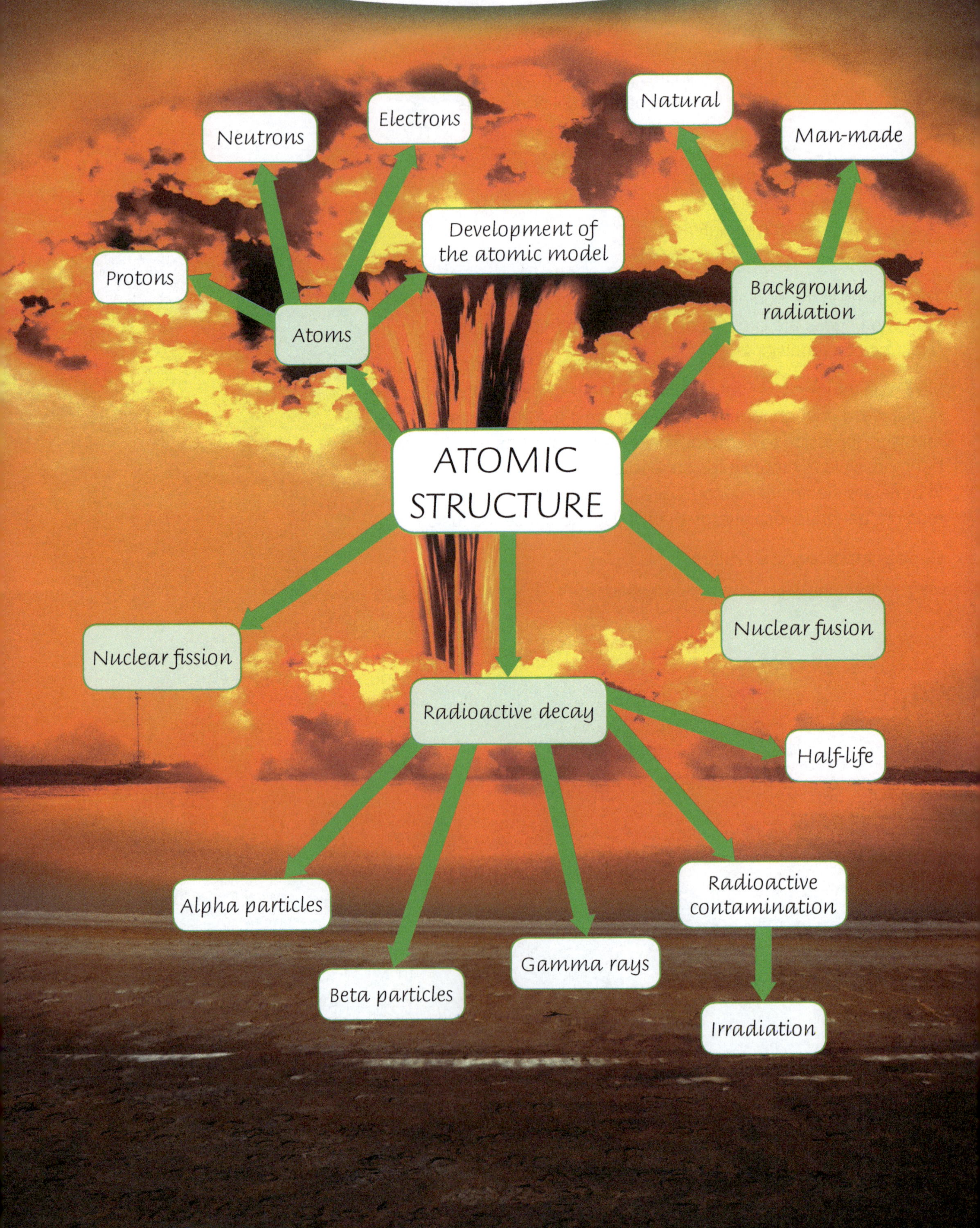

Neutrons

Electrons

Natural

Man-made

Protons

Development of the atomic model

Atoms

Background radiation

ATOMIC STRUCTURE

Nuclear fission

Nuclear fusion

Radioactive decay

Half-life

Alpha particles

Radioactive contamination

Gamma rays

Beta particles

Irradiation

Practice questions

1. The symbol below shows an element.

$$^{48}_{22}\text{Ti}$$

   a) What is its . . .

   i)   number of protons? **(1 mark)**

   ii) number of electrons? **(1 mark)**

   iii) number of neutrons? **(1 mark)**

   b) For parts **i)–iii)** above, explain how you arrived at your answers. **(6 marks)**

2. Experiments by Rutherford, Geiger and Marsden led to the plum pudding model being replaced by the nuclear model.

   a) Explain the differences between the plum pudding model and the nuclear model. **(2 marks)**

   b) How did the further work of Bohr and the further work of Chadwick refine the nuclear model? **(2 marks)**

3. The graph below shows the activity of a radioactive sample over time.

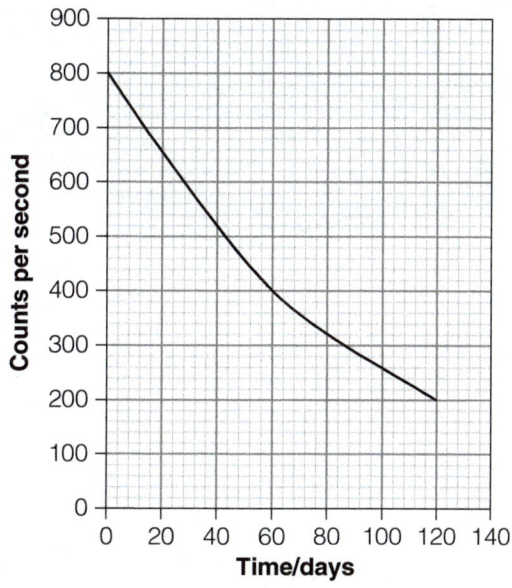

   a) What is the half-life of this sample?
   Explain how you arrived at your answer. **(2 marks)**

   b) Would this sample be useful as a radioactive tracer in medicine?
   Explain your answer. **(2 marks)**

   c) Balance the equation below to show the alpha decay of uranium.

$$^{238}_{92}\text{U} \rightarrow {}^{234}_{90}\text{Th} + {}^{4}_{2}\text{H}$$

   (2 marks)

Our solar system and the life cycle of a star

The planets in the **solar system** orbit the Sun (a **star**). Dwarf planets, such as Pluto, which also orbit the Sun and the natural satellites of planets (moons), are also part of the solar system.

NEPTUNE

URANUS

SATURN

JUPITER

MARS

EARTH

VENUS

MERCURY

Not to scale

The Sun

The Sun was formed from dust and gas (nebula) pulled together by gravitational attraction. Collisions between particles caused the temperature to increase enough for hydrogen nuclei to fuse together forming helium.

The energy released by nuclear fusion processes keeps the core of the Sun hot.

The Sun is stable. The force of gravity acting inwards and trying to collapse the Sun is in equilibrium (balanced) with the outward force produced by the fusion energy trying to expand the Sun.

The Sun is in the main sequence period of its life cycle.

The life cycle of a star

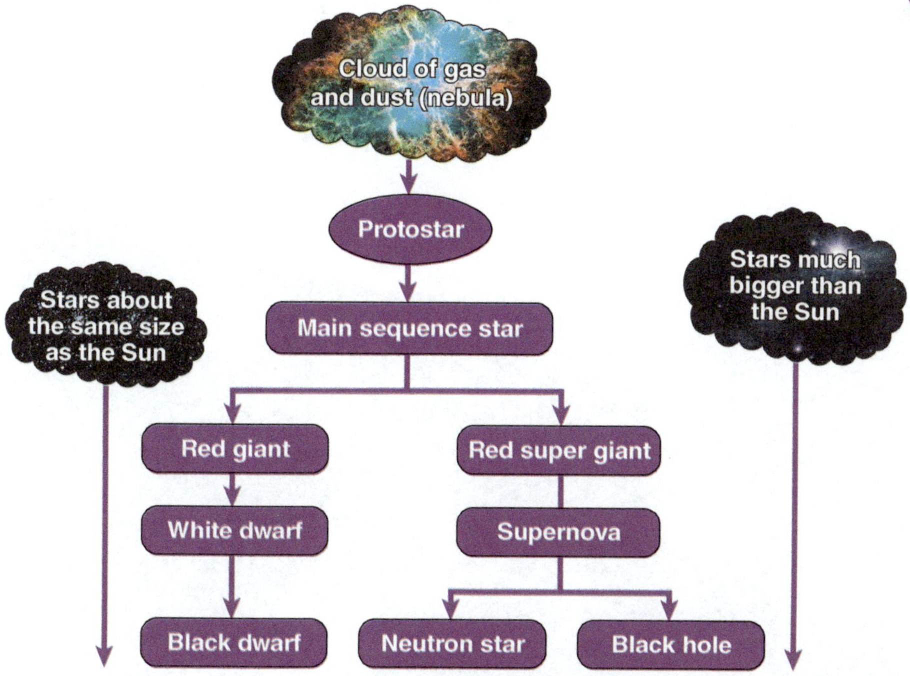

Cloud of gas and dust (nebula)

↓

Protostar

↓

Main sequence star

Stars about the same size as the Sun

Stars much bigger than the Sun

Red giant → White dwarf → Black dwarf

Red super giant → Supernova → Neutron star / Black hole

Keywords

Solar system ➤ Bodies orbiting a star

Star ➤ Very large body of gas held together by its own gravity

➤ Fusion processes in stars produce all of the naturally occurring elements.

➤ All elements heavier than iron are produced in a supernova.

➤ The explosion of a supernova distributes the elements throughout the Universe.

Write each of the stages of the life cycle of a star on separate cards. Then rearrange the cards to show the life cycle of a star that is the same size of the Sun, and one that is bigger.

1. What stage in its life cycle is the Sun currently in?
2. In a stable star, what force balances the fusion trying to expand the Sun?
3. What determines if a star forms a supernova or a white dwarf?
4. How are elements produced that are heavier than iron?

Orbital motion, natural and artificial satellites

Natural satellites

Planets orbit the Sun and a moon orbits a planet.

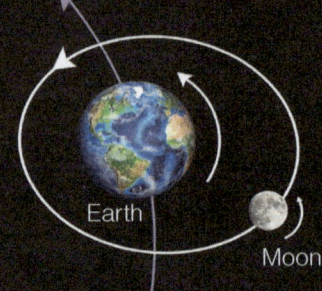

Earth

Moon

Sun

Not to scale

Man-made satellites

Artificial (man-made) **satellites** orbit the Earth and have a variety of functions including communication and weather monitoring.

Satellites in **geostationary orbits** always stay above the same point of the Earth. Satellites in **polar orbits** travel over both poles during an orbit. These orbits often only take a few hours.

Gravity provides the force that allows planets and satellites to maintain their circular orbits.

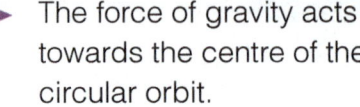

HT

➤ The force of gravity acts towards the centre of the circular orbit.
➤ This causes the object to accelerate in that direction.
➤ The instantaneous velocity of the orbiting body is at a right angle to the direction of the force of gravity.
➤ The speed of the orbiting body is constant. If the body changes speed the orbital height will change.

Red shift

Our solar system is a small part of the **Milky Way** galaxy. There are hundreds of billions of other galaxies in the Universe.

Keywords

Satellite ➤ Object in orbit around a body
Gravity ➤ Process where all objects with mass attract each other

There is an increase in the wavelength of light from most distant galaxies. As the wavelength of visible light increases this leads to the light shifting to the red end of the spectrum. This effect is called **red-shift**.

Galaxies are moving away from the Earth. The further away the galaxies, the faster they are moving. Observations of supernovae suggest that distant galaxies are receding ever faster.

The faster movement leads to a larger observed increase in wavelength and so a larger red-shift.

WS This observed red-shift provides evidence that the Universe is expanding and also supports the **Big Bang theory**. The Big Bang theory states that the Universe began from a very small region that was extremely hot and dense. Cosmic background radiation (**CMB**) also provides evidence for the Big Bang theory.

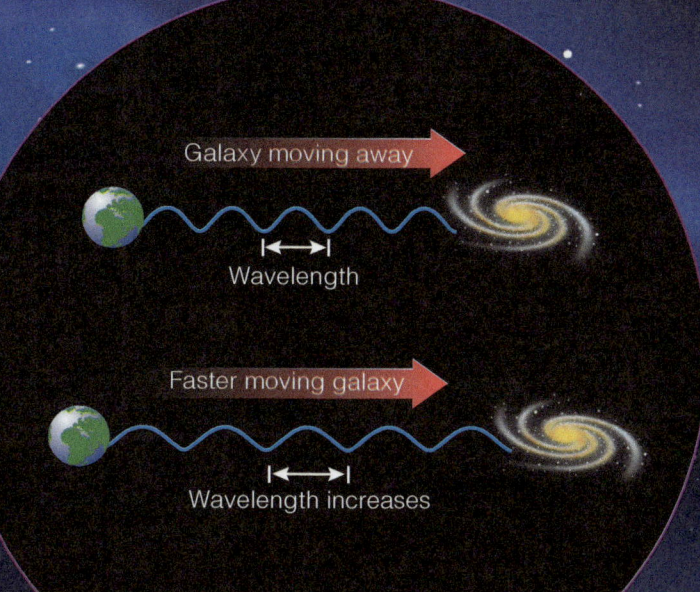

Galaxy moving away

Wavelength

Faster moving galaxy

Wavelength increases

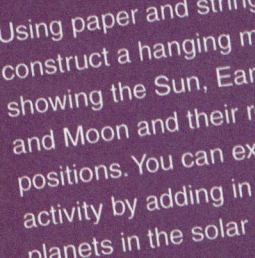

Using paper and string, construct a hanging mobile showing the Sun, Earth and Moon and their relative positions. You can extend this activity by adding in the other planets in the solar system.

1. Give two examples of uses of artificial satellites.
2. What happens to the speed and velocity of a moon as it orbits a planet?
3. What happens to the wavelength of light from a galaxy that is moving away from the Earth?
4. How does red-shift provide evidence for the Big Bang theory?

Mind map

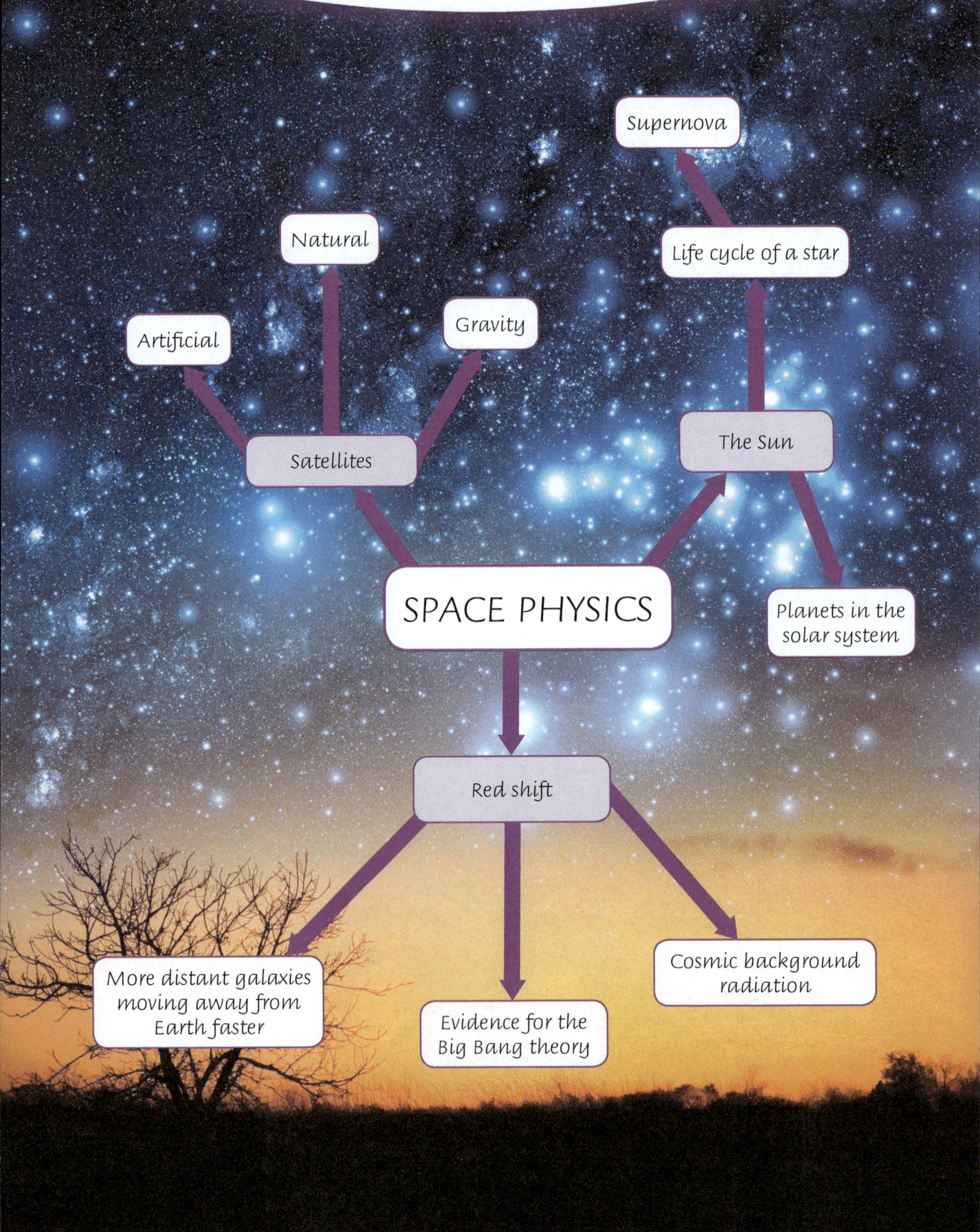

Supernova

Natural

Artificial

Gravity

Life cycle of a star

Satellites

The Sun

SPACE PHYSICS

Planets in the solar system

Red shift

More distant galaxies moving away from Earth faster

Evidence for the Big Bang theory

Cosmic background radiation

Practice questions

1. **a)** Name two examples of natural satellites. **(1 mark)**

   b) Use the concept of balanced forces to explain why the Sun is currently stable. **(3 marks)**

   c) How will the final fate of the Sun differ from a star that has a much larger mass than the Sun? **(3 marks)**

2. **a)** The table below shows the changes in the observed wavelengths of light from three galaxies.

Galaxy	Change in wavelength
1	Increase
2	Decrease
3	Increase

   i) Which of the galaxies' light is red-shifted? **(1 mark)**

   ii) Explain how you arrived at your answer to part **i)**. **(1 mark)**

   iii) What statement can be made about the movement of the galaxies whose light is red-shifted? **(1 mark)**

   iv) What theory of the beginning of the Universe does this help support? **(1 mark)**

   (b) Light from galaxies can also be blue-shifted.
   Explain what you think is meant by this term. **(3 marks)**

3. **a) i)** Draw a line to match the type of satellite to its most likely orbit. **(2 marks)**

   | Communication satellite | | Polar orbit |

   | Weather satellite | | Geostationary orbit |

   ii) Give a reason for each of your answers. **(2 marks)**

   b) i) When a satellite is in a stable circular orbit, what statement can be made about the velocity of the satellite? **(1 mark)**

   ii) What effect would a change in speed have on the orbital radius of a satellite? **(1 mark)**

Answers

Forces 1

Page 5

1. A scalar quantity only has a magnitude, a vector quantity has both a magnitude and a direction.
2. **Two from**: friction; air resistance; tension; normal contact force
3. $0.067 \times 10 = 0.67$ N
4. 78 J

Page 7

1. $0.1 \times 2 = 0.2$ N
2. Any point where the limit of proportionality hasn't been exceeded
3. Elastic potential energy

Page 9

1. The turning effect of a force
2. The total clockwise moment about the pivot equals the total anticlockwise moment about the pivot
3. $72 \times 0.7 = 50.4$ Nm

Page 11

1. $13 \times 1000 \times 10 = 130\,000$ Pa = 130 kPa
2. An object sinks because its weight is greater than the upthrust of the water
3. As the height of an object increases there are fewer air molecules above the object so their total weight is smaller.

Page 13

1. Speed has a magnitude but not a direction.
2. $\frac{10}{2.5} = 4$ km/h
3. Speed is a scalar quantity as it only has a magnitude. Velocity is speed in a given direction. Velocity has a magnitude and a direction so it is a vector quantity.

Page 15

1. Speed
2. A negative acceleration
3. The distance the object travels

Page 17

1. a) weight = mass × gravitational field strength
   $3.7 \times 187 = 691.9$ N **(2 marks)**
   b) Yes. The rover would weigh more (1870 N) **(1 mark)** as the gravitational field strength on Earth is greater than that on Mars **(1 mark)**
   c) pressure = height of the column × density of the liquid × gravitational field strength
   $= 3 \times 1000 \times 3.7 = 11.1$ kPa **(2 marks)**
2. a) i) Gradient of graph from 0–3 s $= \frac{25}{3} = 8.33$
      Acceleration = 8.33 m/s² **(2 marks)**
      ii) Area under graph 0–3 s $= 0.5 \times (3 \times 25) = 37.5$
      Distance travelled = 37.5 m **(2 marks)**
   b) Yes **(1 mark)** as the line on the graph is steeper from 0–3 than 3–6 / A steeper line indicates a faster acceleration **(1 mark)**

Forces 2

Page 19

1. 67 N
2. $89 \times 10 = 890$ N
3. It continues to move at the same speed.

Page 21

1. Thinking distance and braking distance
2. **Two from**: rain; ice; snow
3. It may lead to brakes overheating and/or loss of control.

Page 23

1. Airbags slow down a person's change in momentum in a collision.
2. $6 \times 4 = 24$ kg m/s
3. The momentum changes.

Page 25

1. a) The upward force from the ground is 890 N **(1 mark)** as it is equal and opposite to the force the cyclist is exerting on the ground **(1 mark)**.
   b) As the forces on the rider are balanced **(1 mark)** there is no resultant force and the rider would continue to move at a constant speed. This is Newton's first law **(1 mark)**.
   c) The resultant force would be smaller **(1 mark)** so the rider would slow down **(1 mark)**.
2. a) resultant force = mass × acceleration
      $= 1200 \times 2.5 = 3000$ N or 3 kN **(2 marks)**
   b) momentum = mass × velocity
      $= 1200 \times 16 = 19\,200$ N kg m/s
   c) $F = \frac{m \Delta v}{\Delta t}$ **(1 mark)**
      momentum after crash = 0
      change in momentum = 19 200 **(1 mark)**
      force = 19 200 ÷ 0.4
      force = 48 000 N or 48 kN **(1 mark)**
   d) **Two from**: airbags; crumple zones; seatbelts **(2 marks)**

Energy

Page 27

1. $0.5 \times 0.065 \times 6^2 = 11.7$ J
2. $17 \times 987 \times 10 = 167.79$ kJ
3. It is the amount of energy required to raise the temperature of one kilogram of a substance by one degree Celsius.

Page 29

1. To ensure more energy is usefully transferred and less is wasted
2. $\frac{400}{652} = 0.61 = 61\%$
3. It reduces the thermal conductivity.

Page 31

1. Renewable. **Three from**: bio-fuel; wind; hydro-electricity; geothermal; tidal power; solar power; water waves
   Non-renewable. **Three from**: coal; oil; gas; nuclear fuel
2. The wind doesn't always blow and it's not always sunny, so electricity isn't always generated.
3. They produce carbon dioxide, which is a greenhouse gas. Increased greenhouse gas emissions are leading to climate change. Particulates and other pollutants are also released, which cause respiratory problems.

Page 33

1. a) Renewable resources can be replenished as they are used **(1 mark)** whilst non-renewable energy resources will eventually run out **(1 mark)**
   b) Accidents at nuclear power stations could have devastating health and environmental effects **(1 mark)**. Nuclear waste is very hazardous **(1 mark)**.
   c) **Two from**: building tidal power stations involves the destruction of habitats; wind energy doesn't produce electricity all the time; some people think wind turbines are ugly and spoil the landscape **(2 marks)**
2. a) kinetic energy = 0.5 × mass × (speed)²
       = 0.5 × 1600 × 32²
       = 819 200 N or 819.2 kN **(2 marks)**
   b) efficiency = $\frac{\text{useful output energy transfer}}{\text{useful input energy transfer}}$
       = $\frac{819.2}{1500}$ = 0.55 or 55% **(2 marks)**
   c) Sound **(1 mark)** and heat **(1 mark)**
   d) **One from**: lubrication; oil **(1 mark)**
3. a) g.p.e. = mass × gravitational field strength × height
       = 77 × 150 × 10
       = 115.5 kN **(2 marks)**
   b) kinetic energy = 0.5 × mass × (speed)²
       = 0.5 × 77 × 20²
       = 15.4 kN **(2 marks)**
   c) Gravitational potential energy to kinetic energy **(1 mark)** to elastic potential energy **(1 mark)**

Waves

Page 35

1. Longitudinal: Sound wave
   Transverse: **One from**: water wave; electromagnetic wave
2. The distance from a point on one wave to the equivalent point on an adjacent wave
3. It will increase

Page 37

1. The angle of reflection
2. Ear drum
3. 20 Hz to 20 kHz

Page 39

1. P-waves are longitudinal and travel at different speeds through solids and liquids. S-waves are transverse and cannot travel through a liquid.
2. Similarity: Both use reflection of waves.
   Difference: In ultrasound, waves are partially reflected when they meet a boundary between two different media. In echo sound, waves are reflected off the seafloor or solid objects in the water.
3. They cannot travel through the liquid core of the Earth.

Page 41

1. Infrared radiation
2. Visible light
3. Towards the normal

Page 43

1. By oscillations in electrical circuits
2. Mutation of genes and cancer
3. **One from**: energy efficient lamps; sun tanning

Page 45

1. A virtual image
2. Blue
3. 4.25 cm

Page 47

1. It absorbs all of the radiation incident on it, it does not reflect or transmit any radiation and it is the best possible emitter as it emits the maximum amount of radiation possible at a given temperature.

2. **Two from**: the rate of absorption of radiation; the rate of emission of radiation; the rate of reflection of radiation into space
3. Its temperature will remain constant

Page 49

1. Compression is a region in a longitudinal wave where the particles are closer together **(1 mark)**. Rarefaction is a region in a longitudinal wave where the particles are further apart **(1 mark)**
2. frequency = $\frac{\text{wave speed}}{\text{wavelength}}$
       = $\frac{200}{3.2}$
       = 62.5 Hz **(2 marks)**
3. P-waves travel through solids and liquids **(1 mark)** whereas S-waves cannot travel through a liquid **(1 mark)**. Only P-waves are detected on the opposite side of the Earth from the epicentre of the earthquake, not S-waves **(1 mark)** therefore the outer core of the Earth is liquid **(1 mark)**.
4. original height = $\frac{\text{image height}}{\text{magnification}}$
       = $\frac{5}{2.2}$
       = 2.3 cm **(2 marks)**
5. a) Convex lens **(1 mark)**
   b)

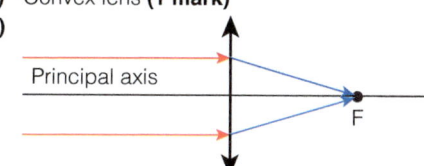

Principal axis

F

(1 mark)

6. Green plants reflect green light so appear green **(1 mark)** whilst all other colours of light are absorbed by the plant **(1 mark)**.

7. Sound travels as a **longitudinal** wave. When a sound wave travels in air the oscillations of the air particles are **parallel** to the direction of energy transfer. The sound wave **doesn't transfer matter, only energy**. **(3 marks)**

8. Ultrasound – sounds with a frequency over 20 kHz **(1 mark)**
   Infrasound – sounds with a frequency less than 20 Hz **(1 mark)**

9. High frequency sound (sonar) is produced by the fishing boat **(1 mark)**. The high frequency sound is reflected off the shoal of fish and the reflection is detected by the boat **(1 mark)**. The difference in time between the ultrasound being produced and the reflection being detected allows the distance to the fish shoal to be determined **(1 mark)**.

Electricity

Page 51

1. The symbol for a variable resistor has an arrow through it.
2. 4.2 C
3. 90 A

Page 53

1. Because resistance depends on light intensity
2. An ammeter and a voltmeter
3. 8 V
4. 2 Ω

Page 55

1. 5 A
2. Add the resistances of both resistors together
3. Lower

Page 57

1. An earth connection is not required as it is impossible for the case to become live.
2. Circuit breakers can be reset and operate faster than a fuse.
3. Touching a live wire would produce a large potential difference across the body.

Page 59

1. $5^2 \times 2 = 50$ W
2. $60 \times 12 = 720$ J
3. They lower the potential difference of the transmission cables to a safe level for domestic use.

Page 61

1. The loss of electrons
2. They will attract
3. To prevent a spark from static electricity, which could ignite the fuel.

Page 63

1. a) current = $\dfrac{\text{potential difference}}{\text{resistance}}$

   $= \dfrac{15}{3}$

   $= 5$ A **(3 marks)**

   b) 5 A **(1 mark)**. The current is the same at all points in a series circuit **(1 mark)**.

   c)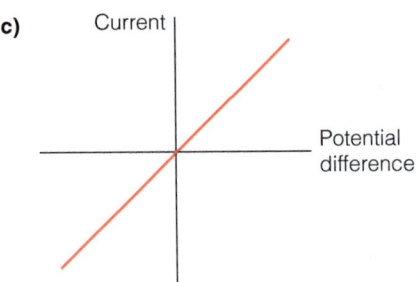

   (3 marks)

   d) **One from**: filament lamp; LDR; diode; thermistor **(1 mark)**

2. a) The bits of paper were either uncharged or negatively charged **(1 mark)** because they were attracted to the positively charged rod **(1 mark)**.

   b) Electrons were transferred from the rod to the cloth **(1 mark)**. The loss of electrons meant the rod had a positive charge **(1 mark)**.

   c) They would repel **(1 mark)** as they have the same charge (positive) **(1 mark)**.

Magnetism and Electromagnetism

Page 65

1. They would repel
2. When it's placed in a magnetic field
3. At the poles of the magnet
4. The Earth's core is magnetic and produces a magnetic field.

Page 67

1. The current through the wire and the distance from the wire
2. The same shape as the magnetic field around a bar magnet
3. Magnetic field, force and current

Page 69

1. It is easily magnetised
2. direct current
3. By increasing the speed of movement or increasing the strength of the magnetic field

Page 71

1. a)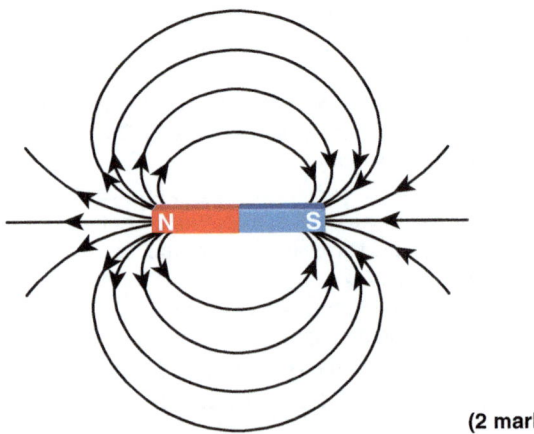

   (2 marks)

   b) The magnets would attract **(1 mark)**.

   c) An induced magnet is only magnetic when placed in a magnetic field whilst a bar magnet is always magnetic **or** An induced magnet always experiences a force of attraction while a bar magnet can experience a force of attraction or repulsion. **(2 marks)**

2. It is a step-down transformer **(1 mark)** as there are fewer coils on the secondary coil than on the primary coil **(1 mark)**. A step-up transformer would have more coils on the secondary coil than the primary coil **(1 mark)**.

3. Step-up transformers increase voltage for transmission in high voltage cables **(1 mark)**. The current is kept low, reducing loss of heat energy **(1 mark)**. A step-down transformer lowers the voltage to a safer level for use in houses **(1 mark)**.

4. $\dfrac{\text{number of turns on primary coil}}{\text{number of turns on secondary coil}} = \dfrac{6}{2} = 3$ **(1 mark)**

   $3 = \dfrac{\text{potential difference on primary coil}}{\text{potential difference on secondary coil}}$

   $3 = \dfrac{\text{potential difference on primary coil}}{15 \text{ V}}$

   $3 \times 15 = $ potential difference on primary coil
   potential difference on primary coil = 45 V **(1 mark)**

Particle Model of Matter

Page 73

1. $200 \times 126 = 25\ 200$ kJ or 25.2 ms
2. Specific latent heat of fusion is the energy required for a change of state from solid to liquid. Specific latent heat of vapourisation is the energy required for a change of state from liquid to vapour.
3. It is changing state.

Page 75

1. When the molecules collide with the wall of their container they exert a force on the wall, causing pressure.
2. 305 Kelvin
3. The internal energy of the gas increases.

Page 77

1. a) density = $\dfrac{\text{mass}}{\text{volume}}$

   $= \dfrac{0.15}{0.0001}$

   $= 1500$ kg/m³ **(2 marks)**

   b) The mass remains constant **(1 mark)** and the density decreases **(1 mark)**.

   c) The temperature would remain constant as it was changing state **(1 mark)** and the solid's internal energy would increase **(1 mark)**.

   d) energy for a change of state = mass × specific latent heat

   $= 0.15 \times 574\ 000$

   $= 86.1$ kJ or 86100 J **(2 marks)**

2. a) i) The particles gain kinetic energy as the temperature increases **(1 mark)**. This increases the speed at which the particles collide with the sides of the container they are in **(1 mark)**.

ii) The same number of particles occupy a greater volume **(1 mark)**, reducing the rate of collisions with the sides of the container **(1 mark)**.

b) $p_1 \times V_1 = p_2 \times V_2$
$500 \times 2.1 = 134 \times V_2$ **(1 mark)**
$\frac{(500 \times 2.1)}{134} = V_2$ **(1 mark)**
$V_2 = 7.8 \text{ m}^3$ **(1 mark)**

3. a) $300 - 273 = 27\ °C$ **(1 mark)**

b) The average kinetic energy and average speed of the particles in the solid would both decrease **(1 mark)**.

4. Specific latent heat of vaporisation **(1 mark)**

Atomic Structure

Page 79

1. The electrons become excited and move to a higher energy level, further from the nucleus.

2. Isotopes

3. The plum pudding model suggested the atom was a ball of positive charge with negative electrons embedded in it. The nuclear model has a nucleus with electrons orbiting.

Page 81

1. A few centimetres

2. **Accept any dense material**, e.g.: concrete or lead.

3. The mass of the nucleus doesn't change but the charge of the nucleus does change.

Page 83

1. When the time taken for the number of nuclei in a sample of the isotope halves, or when the count rate (or activity) from a sample containing the isotope falls to half of its initial level.

2. 5 months

3. The radioactive atoms decay and release radiation.

Page 85

1. Man-made source. **One from**: fallout from nuclear weapons testing; fallout from nuclear accidents
Natural source. **One from**: some rocks (such as granite); cosmic rays from space

2. A short half-life

3. **One from**: killing cells (such as cancer); exploring internal organs (such as tracers)

Page 87

1. Further fission or a chain reaction

2. The energy release from a nuclear reactor is controlled.
The energy release from a nuclear weapon is uncontrolled.

3. To overcome the electrostatic repulsion between the protons in the nucleus

Page 89

1. a) i) 22 **(1 mark)**
ii) 22 **(1 mark)**
iii) 26 **(1 mark)**

b) i) Number of protons is the atomic number **(1 mark)**, which is the bottom number on the symbol **(1 mark)**.

ii) The number of electrons equals the number of protons **(1 mark)**. The number of protons is 28 **(1 mark)**.

iii) The number of neutrons is the mass number **(1 mark)**. The number of protons is 48 – 22 = 26 **(1 mark)**.

2. a) The plum pudding model suggested that the atom is a ball of positive charge with negative electrons embedded in it **(1 mark)**. Rutherford, Geiger and Marsden said the mass of an atom was concentrated in the nucleus with electrons orbiting the nucleus **(1 mark)**.

b) Niels Bohr suggested that the electrons orbit the nucleus at specific distances **(1 mark)**. Chadwick provided the evidence for the existence of the neutron within the nucleus **(1 mark)**.

3. a) 60 days **(1 mark)**. This is the point where the counts per second have halved (Also accept $\frac{800}{2} = 400$) **(1 mark)**

b) No **(1 mark)** because its half-life is too long so it would persist in the body for too long **(1 mark)**.

c) $^{238}U \rightarrow\ ^{234}Th +\ ^{4}He$ **(2 marks)**

Space Physics

Page 91

1. Main sequence

2. The force of gravity

3. Its mass

4. In a supernova

Page 93

1. **Two from**: weather monitoring; spying; communications

2. The speed remains constant. The velocity constantly changes.

3. The wavelength increases and is shifted to the red end of the spectrum.

4. Galaxies are moving away from the Earth. The further away the galaxies, the faster they are moving.

Page 95

1. a) **Two from**: moons; planets; comets **(1 mark)**

b) The outward force from nuclear fusion **(1 mark)** is balanced by **(1 mark)** the force of gravity **(1 mark)**.

c) The Sun will become a black dwarf **(1 mark)**. A larger star would become a neutron star or a black hole **(1 mark)** and would supernova while our Sun wouldn't **(1 mark)**.

2. a) i) Galaxies 1 and 3 **(1 mark)**
ii) These wavelengths increase so the light is moving towards the red end of the spectrum **(1 mark)**.
iii) These galaxies are moving away from Earth **(1 mark)**.
iv) The Big Bang theory **(1 mark)**

b) The wavelength of light decreases **(1 mark)** so is shifted towards the blue end of the spectrum **(1 mark)** so the galaxies are moving towards Earth **(1 mark)**.

3. a) i)

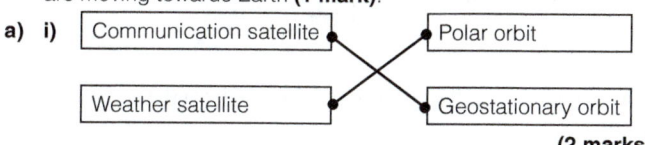

(2 marks)

ii) Communication satellite – needs to stay above the same point on the Earth **(1 mark)**
Weather satellite – travels over both poles in an orbit and can track weather systems **(1 mark)**

b) i) The velocity is constantly changing **(1 mark)**.
ii) A change in speed would cause the orbital radius to change **(1 mark)**.

Physics Equations

force = mass × acceleration

kinetic energy = 0.5 × mass × (speed)^2

momentum = mass × velocity

work done = force × distance (along the line of action of the force)

$$\text{power} = \frac{\text{work done}}{\text{time}}$$

$$\text{efficiency} = \frac{\text{useful output energy transfer}}{\text{useful input energy transfer}}$$

gravity force = mass × gravity constant (g)

potential energy = mass × height × gravitational constant (g) in a gravitational field

force exerted by a spring = extension × spring constant

moment of a force = force × distance (normal to direction of the force)

distance travelled = speed × time

$$\text{acceleration} = \frac{\text{change in velocity}}{\text{time}}$$

wave speed = frequency × wavelength

charge flow = current × time

potential difference = current × resistance

power = potential difference × current = (current)^2 × resistance

energy transferred = power × time = charge flow × potential difference

Physics Equations

$$density = \frac{mass}{volume}$$

$$pressure = \frac{\text{force normal to a surface}}{\text{area of that surface}}$$

$$(\text{final velocity})^2 - (\text{initial velocity})^2 = 2 \times acceleration \times distance$$

change in thermal energy = m × specific heat capacity × change in temperature

thermal energy for a change of state = m × specific latent heat

energy transferred in stretching = 0.5 × spring constant × $(\text{extension})^2$

$$\frac{\text{potential difference across primary coil}}{\times \text{ current in primary coil}} = \frac{\text{potential difference across secondary coil}}{\times \text{ current in secondary coil}}$$

pressure × volume = constant (for a given mass of gas and at a constant temperature) **for gases**

force on a conductor (at right angles to a magnetic field) carrying a current = magnetic flux density × current × length

$$\frac{\text{potential difference across primary coil}}{\text{potential difference across secondary coil}} = \frac{\text{number of turns in primary coil}}{\text{number of turns in secondary coil}}$$

pressure due to a column of liquid = height of column × density of liquid × g

The Periodic Table

Key

| Metals |
| Non-metals |

Relative atomic mass	→	1
Atomic symbol	→	**H**
Name	→	hydrogen
Atomic number	→	1

Group	1	2											3	4	5	6	7	0 or 8
																		4 **He** helium 2
	7 **Li** lithium 3	9 **Be** beryllium 4											11 **B** boron 5	12 **C** carbon 6	14 **N** nitrogen 7	16 **O** oxygen 8	19 **F** fluorine 9	20 **Ne** neon 10
	23 **Na** sodium 11	24 **Mg** magnesium 12											27 **Al** aluminium 13	28 **Si** silicon 14	31 **P** phosphorus 15	32 **S** sulfur 16	35.5 **Cl** chlorine 17	40 **Ar** argon 18
	39 **K** potassium 19	40 **Ca** calcium 20	45 **Sc** scandium 21	48 **Ti** titanium 22	51 **V** vanadium 23	52 **Cr** chromium 24	55 **Mn** manganese 25	56 **Fe** iron 26	59 **Co** cobalt 27	59 **Ni** nickel 28	63.5 **Cu** copper 29	65 **Zn** zinc 30	70 **Ga** gallium 31	73 **Ge** germanium 32	75 **As** arsenic 33	79 **Se** selenium 34	80 **Br** bromine 35	84 **Kr** krypton 36
	85 **Rb** rubidium 37	88 **Sr** strontium 38	89 **Y** yttrium 39	91 **Zr** zirconium 40	93 **Nb** niobium 41	96 **Mo** molybdenum 42	[98] **Tc** technetium 43	101 **Ru** ruthenium 44	103 **Rh** rhodium 45	106 **Pd** palladium 46	108 **Ag** silver 47	112 **Cd** cadmium 48	115 **In** indium 49	119 **Sn** tin 50	122 **Sb** antimony 51	128 **Te** tellurium 52	127 **I** iodine 53	131 **Xe** xenon 54
	133 **Cs** caesium 55	137 **Ba** barium 56	139 **La*** lanthanum 57	178 **Hf** hafnium 72	181 **Ta** tantalum 73	184 **W** tungsten 74	186 **Re** rhenium 75	190 **Os** osmium 76	192 **Ir** iridium 77	195 **Pt** platinum 78	197 **Au** gold 79	201 **Hg** mercury 80	204 **Tl** thallium 81	207 **Pb** lead 82	209 **Bi** bismuth 83	[209] **Po** polonium 84	[210] **At** astatine 85	[222] **Rn** radon 86
	[223] **Fr** francium 87	[226] **Ra** radium 88	[227] **Ac*** actinium 89	[261] **Rf** rutherfordium 104	[262] **Db** dubnium 105	[266] **Sg** seaborgium 106	[264] **Bh** bohrium 107	[277] **Hs** hassium 108	[268] **Mt** meitnerium 109	[271] **Ds** darmstadium 110	[272] **Rg** roentgenium 111							

Elements with atomic numbers 112–116 have been reported but not fully authenticated.

*The lanthanoids (atomic numbers 58–71) and the actinoids (atomic numbers 90–103) have been omitted.
The relative atomic masses of copper and chlorine have not been rounded to the nearest whole number.

Notes

Index

ACKNOWLEDGEMENTS

The author and publisher are grateful to the copyright holders for permission to use quoted materials and images.

All images and illustrations are
© Shutterstock.com and
© HarperCollins*Publishers*

Every effort has been made to trace copyright holders and obtain their permission for the use of copyright material. The author and publisher will gladly receive information enabling them to rectify any error or omission in subsequent editions. All facts are correct at time of going to press.

Published by Letts Educational
An imprint of HarperCollins*Publishers*
1 London Bridge Street
London SE1 9GF

ISBN: 9780008318338

Content first published 2016
This edition published 2019

10 9 8 7 6 5 4 3 2 1

© HarperCollins*Publishers* Limited 2019

All rights reserved. No part of this publication may be reproduced, stored in a retrieval system, or transmitted, in any form or by any means, electronic, mechanical, photocopying, recording or otherwise, without the prior permission of Letts Educational.

British Library Cataloguing in Publication Data.

A CIP record of this book is available from the British Library.

Series Concept: Emily Linnett
Author: Dan Foulder
Commissioning and Series Editor: Charlotte Christensen
Editor and Project Manager: Rebecca Rothwell
Inside Concept Design: Paul Oates
Cover Design: Sarah Duxbury
Text Design, Layout and Artwork: Q2A Media
Production: Karen Nulty
Printed and bound by CPI Group (UK) Ltd, Croydon, CR0 4YY